Frederick Putnam Spalding

A Text-Book on Roads and Pavements

Vol. 3

Frederick Putnam Spalding

A Text-Book on Roads and Pavements
Vol. 3

ISBN/EAN: 9783337420130

Printed in Europe, USA, Canada, Australia, Japan

Cover: Foto ©berggeist007 / pixelio.de

More available books at **www.hansebooks.com**

A TEXT-BOOK

ON

ROADS AND PAVEMENTS.

BY

FRED P. SPALDING,

*Assistant Professor of Civil Engineering in Cornell University,
Member American Society of Civil Engineers.*

FIRST EDITION.
FIRST THOUSAND.

NEW YORK:
JOHN WILEY & SONS,
53 EAST TENTH STREET.
1894.

Copyright, 1894,
BY
FRED P. SPALDING.

PREFACE.

SUCCESSFUL practice in the construction of highways must depend upon correct reasoning from elementary principles in each instance rather than upon following definite rules or methods of construction.

The aim of this book is to give a brief discussion, from an engineering standpoint, of the principles involved in highway work, and to outline the more important systems of construction, with a view to forming a text which may serve as a basis for a systematic study of the subject.

Details and statistics of particular examples have for the most part been excluded as undesirable in a book of this character. Such information is available in many forms for those having the necessary elementary training and experience to enable them to properly use it.

Considerable space has been given to the location and construction of country roads, as seemed proper in view of the present general public interest in the matter, and the probable development of this new field of activity in engineering work. The improvement of our common roads must come through transferring such work to the charge of those who make it a profession, and not through teaching the public how roads should be constructed.

F. P. S.

ITHACA, N. Y., July, 1894.

CONTENTS.

CHAPTER I.

GENERAL CONSIDERATIONS.

	PAGE
Art. 1. Object of Roads	1
2. Resistance to Traction	2
3. Tractive Power of Horses	7
4. Desirability of Various Surfaces	9
5. Economic Value	11
6. Healthfulness	14
7. Safety	16
8. Durability	18

CHAPTER II.

DRAINAGE OF STREETS AND ROADS.

Art. 9. Necessity for Drainage	22
10. Surface Drainage	23
11. Subsurface Drainage	26
12. Kinds of Soil	28
13. Types of Drains	31
14. Culverts	34
15. Water-breaks	42

CHAPTER III.

LOCATION OF COUNTRY ROADS.

Art. 16. Considerations governing Location	43
17. Length of Road	47

v

	PAGE
Art. 18. Rise and Fall	49
19. Rate of Grade	51
20. Examination of Country	53
21. Placing the Line	56
22. Comparison of Routes	59
23. Changing Existing Locations	63

CHAPTER IV.

IMPROVEMENT OF COUNTRY ROADS.

Art. 24. Nature of Improvements	67
25. Earthwork	68
26. Drainage	72
27. Earth-road Surface	73
28. Gravel Roads	76
29. Maintenance of Country Roads	78
30. Width of Country Roads	80
31. Economic Value of Road Improvements	82
32. Systems of Road Management	85

CHAPTER V.

BROKEN-STONE ROADS.

Art. 33. Definition	88
34. Macadam Roads	89
35. Telford Foundations	91
36. Choice of Foundation	93
37. Materials	96
38. Binding Material	101
39. Compacting	103
40. Thickness of Road Covering	105
41. Cross-section	107
42. Maintenance	107

CHAPTER VI.

FOUNDATIONS FOR PAVEMENTS.

Art. 43. Preparation of Road-bed	110
44. Purpose of Foundation	111

CONTENTS. vii

	PAGE
Art. 45. Sand Foundation	112
46. Gravel and Broken Stone	113
47. Concrete	114
48. Brick	116
49. Sand and Plank	117
50. Depth of Foundation	117

CHAPTER VII.

BRICK PAVEMENTS.

Art. 51. Paving Brick	119
52. Tests for Paving Brick	122
53. Foundations	128
54. Construction	129
55. Maintenance	132

CHAPTER VIII.

ASPHALT PAVEMENTS.

Art. 56. Asphaltum	134
57. Rock Asphalt	140
58. Asphalt Blocks	142
59. Foundations	144
60. Construction	146
61. Vulcanite or Distillate Pavement	148
62. Maintenance	149

CHAPTER IX.

WOOD PAVEMENTS.

Art. 63. Wood Blocks	151
64. Foundations	154
65. Construction	156
66. Preservation of Wood	161
67. Maintenance	164
68. Healthfulness	165

CONTENTS.

CHAPTER X.

STONE-BLOCK PAVEMENTS.

	PAGE
Art. 69. Stone for Pavements	168
70. Cobblestone Pavements	170
71. Belgian Blocks	172
72. Granite and Sandstone Blocks	172
73. Construction	173
74. Stone Trackways	176

CHAPTER XI.

CITY STREETS.

Art. 75. Arrangement of City Streets	178
76. Width and Cross-section	183
77. Street Grades	187
78. Street Intersections	189
79. Footways	191
80. Curbs and Gutters	196
81. Crossings	202
82. Street-railway Track	203
83. Trees for Streets	211
84. Alleys	212

A TEXT-BOOK
ON
ROADS AND PAVEMENTS.

CHAPTER I.

GENERAL CONSIDERATIONS.

ART. 1. OBJECT OF ROADS.

THE primary object of a road or street is to provide a way for travel, and for the transportation of goods from one place to another. The facility with which traffic may be conducted over any given road depends upon the resistance offered to the passing of vehicles by the surface or the grades of the road, as well as upon the freedom of movement allowed by the width and form of the roadway. In order that a road may offer the least resistance to traffic, it should have as hard and smooth a surface as possible, while affording a good foothold to horses, and should be so located as to give the most direct route with the least gradients.

The expediency of any proposed road construction or improvement depends upon its desirability as affecting the comfort, convenience, and health of residents of the locality, and also upon its economic value which is largely determined by its cost and durability, as well as upon the facility it gives for the conduct of traffic.

The problem of the highway engineer, in designing works of this character, involves the consideration of these various elements and their proper adjustment to give the best results.

The kinds of road surface most commonly employed are as follows: For the streets of cities and towns, pavements of stone blocks, brick, asphalt, and wood; for suburban streets and important country roads, macadam and gravel surfaces; for ordinary country roads in general, surfaces of earth or gravel.

ART. 2. RESISTANCE TO TRACTION.

The resistance to traction of a vehicle on a road surface may be divided into three parts: axle friction, rolling resistance, and grade resistance.

Axle friction varies with the nature of the bearing surfaces, and for vehicles of similar construction is directly proportional to the load. It is entirely independent of the nature of the road surface.

Rolling resistance is of two kinds: that due to irregularities in the surface of the road, and that of a wheel to rolling upon a smooth surface, sometimes called rolling friction.

The resistance due to an inequality in the road surface, is the horizontal force necessary, at the axle, to raise the weight upon the wheel to the height of the obstacle to be passed. Thus (Fig. 1), by the principle of the lever, $P = W\dfrac{c}{a}$.

For small inequalities, this resistance will be approximately inversely as the diameter of the wheel. The effect of small irregularities in the surface, however, is due more to the shocks and concussions produced by them than to the direct lifting action of the obstacle, and the resistance due to uneven surface is greater at high than at low velocities.

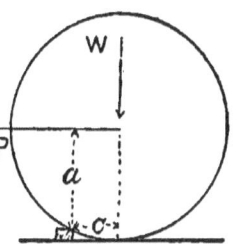

FIG. 1.

Rolling friction is probably due for the most part to the compressibility of the surface of the road, which permits the wheel to indent it to some extent. The wheel is thus always forcing a wave of the surface before it, or climbing an inclination caused by its weight upon the road surface. This rolling friction varies for wheels of differing diameters, being less for large than for small wheels. The experiments of M. Morin, in France, seemed to indicate that the resistance varies inversely as the diameter. Other experiments have indicated a less variation, approximately as the square root of the diameter, while Mr. D. K. Clark (Roads and Streets by Law and Clark; London, 1890) concludes, from a mathematical discussion based upon the assumption that the material of the surface is homogeneous and the pressure proportional to the depth of penetration, that the resistance to traction is inversely as the cube root of the diameter of the wheel.

For practical purposes it may be considered that, for wheels of ordinary sizes used on road vehicles, the rolling resistances are equal to the load multiplied by a coefficient which depends upon the nature and condition of the road surface, although these coefficients are somewhat affected by the sizes of the wheels.

Many experiments have been made for the purpose of determining the tractive force required for a given load upon various road surfaces. The results show somewhat wide variations, as would be expected when the many elements that may affect them are considered. The following table shows a few average results, which will give some idea of the relative resistances of various surfaces and of the advantage to be derived from a smooth and well-kept road surface:

TRACTIVE RESISTANCES ON VARIOUS SURFACES.

Character of Road.	Resistance, Lbs. per ton.		
Earth—ordinary in fair condition.	125	to	140
dry and hard............	60	"	100
Macadam—very good.............	40	"	60
ordinary...............	60	"	80
poor..................	100	"	150
Granite-block pavement—good...	25	"	40
ordinary	50	"	80
Asphalt pavement................	15	"	25
Wooden-block pavement	20	"	30

On earth roads or smooth pavements the tractive force is independent of the velocity; but on rough pavements, where concussions take place, the tractive force increases as the speed increases.

Grade Resistances.—Tractive resistance due to grade is independent of the nature of the road surface, or of the size of the wheels. It is equal to the load multiplied by the sine of the angle made by the grade with the horizontal. Thus in Fig. 2 the tractive force P, due to the grade, is the force necessary to prevent the wheel from rolling down the slope under the action of the weight W, or it is the component of W parallel to the slope ac.

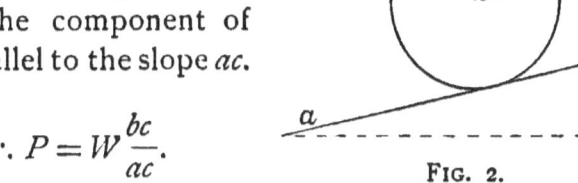

Fig. 2.

$$\therefore P = W\frac{bc}{ac}.$$

Grades are ordinarily expressed in terms of rise or fall in feet per hundred, or as percentage of horizontal distance.

For all ordinary cases of small inclinations ab is approximately equal to ac, and we may take

$$P = W\frac{bc}{ab};$$

or the tractive force necessary to overcome any grade equals the load multiplied by the percentage of grade.

The total tractive force necessary to haul a load up an inclined road equals the sum of the force necessary to haul the load upon the same surface when level, and the force necessary to overcome the grade resistance. Thus, if we wish to find the tractive effort necessary to haul a load of 2 tons up a grade of 3 ft. in 100 over a good macadam road. Taking the resistance of the road surface when level at 60 lbs. per ton, we have for the total resistance

$$R = 2 \times 60 + 4000 \times \tfrac{3}{100} = 240 \text{ lbs.}$$

In going down the grade, the force due to grade becomes a propelling force, and the tractive effort required is the difference between the surface resistance and grade force. In case the grade force be the greater, the resulting tractive force becomes negative, or it will be necessary to apply the force as a resistance to prevent acceleration of the velocity in the descent.

The angle for which the tractive force required for a given surface equals the grade resistance is called the *Angle of Repose* for that surface. In the case given above, $2 \times 60 - 4000 \times \frac{3}{100} = 0$, or the angle of repose for a surface whose level resistance is 60 lbs. per ton is a 3% grade. If a vehicle were left standing upon that inclination, it should remain standing with the forces just balanced. If it were started down the grade, it should continue to move at a uniform rate, without the application of any other force.

In a series of experiments made by the Studebaker Brothers Manufacturing Company (see *The Engineering Record* for Dec. 16, 1893) upon the traction necessary upon various surfaces with American wagons, it was found that the width of wheel-tire has little if any effect upon a hard surface; that there was a small difference in favor of the wide tire upon soft ground, and upon sod the narrow tire would cut through where the wide one would pass over.

The wheels used in the tests had tires $1\frac{3}{4}$, 3, and 4 inches wide, and varied from 3 ft. 6 in. to 4 ft. 6 in. in diameter.

The general results show a variation in tractive force required, depending upon the construction of the vehicle, of from 30 to 65 pounds per ton for a stone-block pavement, from 120 to 175 pounds per ton for a

GENERAL CONSIDERATIONS. 7

good sand road, from 60 to 100 pounds per ton for a gravel road, and from 240 to 325 pounds per ton on a muddy road.

It was also found that the force necessary to start the load was from 125 to 200 pounds per ton greater than that necessary to keep it in motion; the load starting easier on wheels of large diameter than upon small ones, but the diameter seemingly having less effect upon the traction when in motion.

ART. 3. TRACTIVE POWER OF HORSES.

The loads that a horse can pull upon various road surfaces will not necessarily be proportional to the resistance offered by the surface to traction, as the tractive force that the horse can exert depends upon the foothold afforded by the surface. The ability of a horse to exert a tractive force depends upon the strength of the animal, upon his training for the particular work, and whether he be accustomed to the surface upon which he is travelling. The work of different animals is therefore subject to considerable variations, and only very rough approximations are possible in giving average values of the work a horse may do under differing conditions.

The tractive force that may be exerted by a horse, at moderate speeds, varies approximately inversely as the rate of speed; or, in other words, the power that a horse can exert through any considerable time is nearly constant for varying velocities. Thus it may be assumed, as an average value, that a horse working regularly ten hours per day can put forth a tractive effort of 80 pounds at a speed of 250 feet per minute on an ordinary level road surface.

For the power of the horse we then have

$$\text{Power} = \text{force} \times \text{velocity} = 80 \times 250 = 20000 \text{ foot-lbs. per minute.}$$

For any other rate of speed, as 200 feet per minute, we would have $20000 \div 200 = 100$ pounds as the tractive force exerted by the horse.

If the period of daily work be lessened, the power that may be developed will be increased, either by increasing the load or the velocity.

The tractive force that a horse is able to exert decreases very rapidly as the rate of inclination increases. This is due both to the expenditure of power by the horse in lifting his own weight up the grade, and to the less firm footing on the inclination. The effect of differences in the foothold afforded by various pavements is very marked in the loss of tractive power upon grades.

In the table below are given the loads that an average horse may be expected to continuously haul up different inclinations, on various road surfaces, at slow speed. These figures, while of little value as an absolute measure of what may be done in any particular case, are of use as a rough comparison of the relative tractive properties of different surfaces and grades. The effect of grades upon tractive effort will also depend upon the condition in which the surface is maintained, and upon the weather. Snow and ice in winter, or the damp and muddy condition of some pavements in wet weather, have a very considerable effect to diminish tractive power.

LOADS IN POUNDS THAT A HORSE CAN DRAW UPON VARIOUS SURFACES AND GRADES.

Kinds of Surface.	Rate of Grade.							
	Level.	1 in 100.	2 in 100.	3 in 100.	4 in 100.	5 in 100.	10 in 100.	15 in 100.
Earth road—good	3000	2400	2000	1600	1400	1200	800	300
poor	1300	1100	900	700	600	500	400	150
Broken-stone—good	4000	2700	2000	1600	1400	1200	700	200
poor	1600	1100	800	600	500	450	250	100
Stone Blocks—good	6000	4500	3300	2700	2200	1700	900	400
poor	3000	2300	1700	1400	1100	900	450	200
Asphalt—clean and dry	10000	4000	2500	1800	1300	1000	400	....

On steep grades (more than 8 or 10 in 100) special forms of block pavements are sometimes employed to increase the tractive power by affording better foothold to horses. Sheet asphalt is not usually employed on grades of more than 4%. Ordinary wood blocks and brick are used up to grades of 7% or 8%, and granite blocks to 10%.

A horse may frequently exert for a short time a tractive force about double that which he can exert continuously; hence, when short grades occur steeper than the general grades of the road, loads may often be taken over them much heavier than could be carried if the steeper grade prevailed upon the road.

ART. 4. DESIRABILITY OF VARIOUS ROAD SURFACES.

The desirability of a road surface for any particular use depends both upon its fitness for the service required of it and upon its durability in use.

Upon a country road, the problem of improvement ordinarily consists simply in providing the hardest and most durable surface consistent with an economical expenditure of available funds, the object being to lighten the cost of transportation by reducing the resistance to traction, and to render travel easy and comfortable.

Upon city streets, however, several other factors may be of importance in the design of highway improvements.

The comfort both of those using the street and of the occupants of adjoining property will be largely affected by the freedom of the surface from noise and dust.

The safety of the pavement in use, its effect upon the health of residents of the locality, and its economic value must in each case be considered.

To adjust to the best advantage these various elements, frequently quite discordant with each other, is a matter which can only be accomplished by the exercise of good judgment. Local conditions and necessities must always be considered—such as the difficulties of drainage, the availability of various materials, the nature of the traffic to be carried, and the needs of the business or property interests of the neighborhood. Thus, for heavy hauling of a large city, the durability and resistance to wear of the pavement may be the paramount consideration; for an office district, quiet may be very important; for the lighter driving of a residence street, the elements of comfort and healthfulness may properly be considered as of greater force than the purely economic ones; while in all of the cases the necessary limitation of first cost will largely determine what may or may not be done.

Art. 5. Economic Value.

The determination of the economic value of any proposed road or street improvement is always a matter of difficulty, as it embraces so many items which cannot be exactly evaluated. The factors to be considered in this connection are:
1. Cost of construction.
2. Cost of maintenance and repairs.
3. Cost of conducting transportation.
4. Effect upon land values or business interests.

For the purpose of comparing various pavements, or of considering the advisability of any proposed improvement, we may sum the interest on the cost of construction with the annual charges representing the other items, and find which improvement will make the total annual cost a minimum or annual benefit a maximum. This process, in any case in practice, simply amounts to a use of the judgment, having properly in view the various interests to be affected, as to what expense may legitimately be allowed in order to secure a certain benefit. The outlay is usually quite tangible and easily estimated, while the advantages cannot be directly estimated and are often overlooked. It must not, however, be supposed that they have no financial value, or that the pavement which can be constructed and maintained for the least money is necessarily the most economical to use.

The cost of construction must, of course, include everything connected with the original construction of the road and all necessary expenses leading to the improvement under consideration.

The cost of maintenance and repairs includes an estimate of the average cost of keeping the road in good condition over a term of years, taking into account the necessity of renewing the surface at the expiration of the life of the pavement, the cost of cleaning and sprinkling, and such minor repairs as may be necessary from time to time to maintain a uniformly good surface.

An approximate estimate upon these points may usually be made by examining the records of the same kind of construction under similar conditions elsewhere. The cost of maintenance of each kind of pavement varies widely in different localities and under differing treatment, and no general rules can be stated as to the relative costs of the various systems.

All road surfaces will require maintenance, the same as any other class of engineering constructions subjected to wear in use; and as a rule the cost of maintenance will be less as the care used in keeping the surface always clean and in good condition is greater.

The cost of transportation is affected by the nature and condition of the road over which the traffic must pass, both because the resistance to traction offered by the surface determines the load that may be hauled over it, and because the roughness of the surface serves both to limit speed and to cause wear upon horses, harness, and vehicles. The evaluation of these items is a matter of difficulty, on account of the practically indeterminate nature of the data upon which they should be based.

A rough idea of the relative cost of transportation over different road surfaces may sometimes be obtained by observing or estimating the extent and nature of the traffic that is likely to pass over the road and esti-

GENERAL CONSIDERATIONS. 13

mating the cost of carrying this traffic over each surface. The portion of the traffic which consists in hauling maximum loads will be directly affected by differences in tractive resistances; the number of loads necessary to move the traffic, and hence the cost, being, for this portion, approximately proportional to the resistance. For the lighter portion of the traffic the greater speed with a smooth surface and easy grades will be of value in the saving of time, although difficult to state in money values.

The effect of a smooth surface is also very appreciable in the cost of wear and tear upon horses, vehicles, and harness. The value of this item is variously estimated, and probably ranges from one to ten cents per mile travelled.

Earth roads, in good condition, and wood pavements seem most favorable to horses, although asphalt and broken-stone roads are commonly considered most advantageous as to general wear. Brick would not differ greatly from asphalt. On earth roads in poor condition the wear is severe, and on stone blocks it is estimated to be three or four times as great as on asphalt. The financial value of the saving in this wear and tear is difficult to ascertain, but it is undoubtedly sufficient to make it an important item in the cost of highway transportation.

Land Values.—The effect of highway improvements upon the value of adjoining property is dependent upon the nature of the uses to which the property may be put, and the extent to which various characteristics of the road surfaces, such as dust or noise, may affect the occupations or comfort of the occupants. A pavement may thus sometimes have a direct effect upon rental values. In general, however, the effect is difficult to estimate, although it is commonly recognized in the practice of assessing a portion of the cost of improvements against abutting property.

The value of comfort, convenience, safety, and healthfulness to a community, as affected by the condition of their roads and streets, cannot readily be stated in figures; but they have a money value, both in their effect upon the general life and business of the community and in the attraction presented to outside business enterprises or home-seekers.

Art. 6. Healthfulness.

The effect of a pavement upon the health of the residents of its locality will be affected by the tendency of the materials composing it to decay, by its permeability, and by its degree of freedom from noise and dust.

The permeability of a road surface is important on account of the tendency of surface-water and refuse matter to penetrate and saturate it, and thus cause it to become dangerous to health. A continuous sheet pavement is the most desirable in this particular, and a block pavement with open joints the least so. When, however, the joints of a block pavement are properly cemented, the pavement may be made nearly impervious. If the material of which the pavement is composed be permeable, it may gradually become saturated with street refuse, even though the joints be made tight, and where the material is liable to decay it may of itself become obnoxious to health.

Both these objections are raised to the use of wood

pavements, and probably in many cases with justice. This is a matter, however, concerning which authorities differ. The extent of the danger to health involved in the use of wood for pavements in any particular case probably depends largely upon the wood selected for use, and the method of construction adopted. It is at least questionable whether the permeability of the material used for pavements is in practice ever as objectionable on the ground of health as that caused by open-joint construction of block pavement, even though the material of the blocks be impervious to moisture.

The noise made by traffic upon a pavement is important not only because of its effect upon the comfort of the people using it or living adjacent to it, but also because to it are frequently attributed many nervous disorders to which people in some cities are subject.

Stone-block pavements are the most objectionable in this particular, causing a continual roar, due both to the rumbling of wheels over them and the blows of the horses' feet upon them. Upon asphalt the noise is only that due to the horses' feet, giving a sharp, clicking sound. Upon wood the horses produce no appreciable sound; but wheels give a dull rumble, generally considered the least objectionable of any of the noises made by the more common pavements. The noise of wood pavements is diminished by making the joints between blocks small. A brick surface gives a combination of the sounds of wood and asphalt, the clicking being much less sharp than on asphalt, and the rumble less noticeable than on wood. On any block pavement the noise is lessened as the foundation is made more firm and the joints more close and well cemented.

An earth or broken-stone road is usually less noisy than any of the hard pavements.

The giving off of dust by a pavement under the action of traffic is also objectionable on the score of health as well as of comfort. All pavements produce more or less dust, the amount depending more upon the method of construction and care used in forming the surface and filling the joints than upon the material of the pavements. For the most part, however, the presence of dust is dependent rather upon the maintenance, cleaning, and sprinkling of the pavement than upon its nature, and the dirt upon the surface of a hard pavement is usually carried there from the outside and not due to the pavement.

Earth and broken-stone roads wear rapidly, and make dust freely in dry weather, requiring frequent sprinkling and cleaning to keep the road clear of it, and are on this account objectionable for use on the streets of towns under any considerable traffic.

Art. 7. Safety.

The safety of a road surface depends upon the foothold afforded by it to horses under normal conditions, and also upon the degree of slipperiness that it may take in wet weather, or under the influence of ice and snow in winter.

A dry earth road in good condition gives the best and surest foothold, with broken-stone and gravel roads nearly as good.

The relative safety of the various pavements used in city streets is a matter upon which there is considerable difference of opinion amongst authorities. Local

GENERAL CONSIDERATIONS. 17

conditions affect the pavement in this regard to an important degree. The dampness of the climate, the shade from buildings, the cleanliness of the streets, and the prevalence of snow and ice in winter are all important.

Statistics upon the question of relative safety of wood, asphalt, and granite have been collected by Capt. Greene in this country and by Col. Haywood in London, the attempt being made to determine the number of miles travelled by horses upon each kind of pavement to each accident due to slipperiness.

The results of Col. Haywood seem to show that of the three wood is the safest and granite the most dangerous, while the results of Capt. Greene show asphalt to be the best and wood the worst in this particular.

Col. Haywood's observations were all taken on London streets, and are as follows:

	Miles travelled to each fall on		
	Granite.	Asphalt.	Wood.
In dry weather,	78	223	646
" damp "	168	125	193
" wet "	432	192	537
All observations,	132	191	330

The observations were made when dry weather prevailed, and therefore are somewhat unfavorable to granite, which is safest when wet.

Capt. Greene's observations were made in several American cities, and showed the distance travelled to each fall to be, on granite 413 miles, on asphalt 583 miles, and on wood 272 miles. The observations on wood in this series were too few to give a reliable indication, and it is to be observed with regard to all of them that slipperiness is largely affected by the condition in which the surface is maintained, and it is therefore difficult to draw any general conclusions which would fit all cases.

All hard pavements are slippery when muddy and wet, and cleanliness is the necessary condition of safety.

Wood and asphalt, if clean, are least slippery when dry and most so when simply damp. Granite, after the surface becomes worn and polished, is most slippery when dry and least so when wet.

Under a light fall of snow both wood and asphalt become very slippery, and in freezing weather wood sometimes becomes slippery through the freezing of the moisture retained by it.

No statistics are available as to the safety of brick pavements, but it is thought a desirable material in this respect.

It may also be remarked, that the danger of a horse falling upon any pavement depends very largely upon the training of the animal and whether he be accustomed to the particular surface in question.

ART. 8. DURABILITY OF VARIOUS SURFACES.

The durability of a road or pavement is dependent upon so many circumstances connected with local conditions, the nature of the traffic, methods of construction, and efficiency of maintenance, that any comparison of the various kinds of pavement in this respect is difficult and likely to be misleading.

The qualities which especially affect the durability of the road may be partially enumerated as follows:

(1) The hardness and toughness of the material composing the surface, upon which depends the resistance of the surface to the abrading action of the wheels and horses' feet passing over it.

(2) The firmness of the foundation, which serves to distribute the loads over the road-bed, and keep the surface uniform.

(3) The drainage of the road-bed, which can only properly sustain the loads which come upon it when it is dry.

(4) The permeability of the surface, which should form a water-tight covering to serve the purpose of keeping the foundation and road-bed in a dry condition.

(5) The resistance of the materials of the pavement to the disintegrating influences of the atmosphere, and to the action of the weather.

The relative importance of these various factors, in any particular case, depends largely upon the nature and extent of the traffic which is to pass over the pavement.

The amount of traffic to which a street is subjected is usually estimated in terms of tons per foot of width of street, by observing the number of teams passing a given point during certain times, classifying them, and assigning an average value of load to each class. The wear of the surface will naturally be somewhat proportional to the amount of traffic. The life of a pavement is, however, affected by other conditions, and hence cannot always be inferred from the amount of traffic.

Traffic may also be classified according to its nature as heavy or light, depending upon the weight of individual loads which are carried. It is the heavy loads borne upon narrow wheel-tires that do the greatest damage to a pavement, and hence the nature rather than the amount of traffic determines the character of pavement necessary.

Granite blocks, where a firm unyielding foundation is employed, give the hardest and most durable surface of any of the common pavements. This is especially the case under very heavy loads.

Asphalt and brick rank next to stone, and when well constructed are satisfactory under any but the heaviest traffic. The relative durability under wear of brick and asphalt is a matter of doubt, both materials being subject to considerable variations in quality, and showing varying results in different localities, due both to differences in the quality of the material and in the methods of construction.

Wood is less durable and only suitable for comparatively light traffic, unless its other advantages be considered worth the high cost of maintenance under heavier traffic, as has been the case in London, where wood has been largely used under traffic which required its renewal every four or five years.

Broken stone wears rapidly under moderately heavy traffic, and should be employed only on suburban streets or country roads used mainly for light driving or a small amount of traffic.

The durability of any pavement also depends largely upon the system employed for maintaining it, and upon its being kept clean. Cleanliness is specially important with wood, asphalt, and broken stone.

Brick or stone blocks are not so much injured by neglect.

The wear of a pavement also depends largely upon the smoothness of the surface, as the impacts to which the material is subjected are produced by irregularities. So that the most durable material may not always give the greatest resistance to wear.

CHAPTER II.

DRAINAGE OF ROADS AND STREETS.

ART. 9. NECESSITY FOR DRAINAGE.

THE road-bed, usually formed of the natural earth over which the road or pavement is to be constructed, must always carry the loads which come upon the road surface. Where an artificial road surface or pavement is employed, the earth road-bed is protected from the wear of the traffic, and the wheel loads coming upon the surface are distributed over a greater area of the road-bed than if the loads come directly upon the earth itself; but the loads are transferred through the pavement to the road-bed, and not sustained by the pavement as a rigid structure.

The ability of earth to sustain a load depends in a large measure upon the amount of moisture contained by it. Most earths form a good firm foundation so long as they are kept dry, but when wet they lose their sustaining power, becoming soft and incoherent. When softened by moisture the soil may be easily displaced by the settling of the foundation of the road, or forced upward into any interstices that may exist in its superstructure.

In cold climates the drainage of a road is also important because of the danger of injury from freezing. Frost has no disturbing effect upon dry material, and hence is an element of danger only in a road that retains water.

In order, therefore, that the loads may be uniformly sustained, and the surface of the road kept firm and even, it is evidently of first importance that the road-bed be maintained in a dry condition. This may be accomplished by the use of an impervious road covering, by proper underdrainage, or by a combination of the two, as may be necessary in any particular case.

An impervious surface is always desirable, not only as a means of keeping the road-bed dry, but also as a protection to the pavement itself against the disintegrating action of water and of the weather upon the materials of the surface. Such a surface is not, however, always practicable, and other means must often be used to free the road from water.

The necessity for underdrainage in any case depends upon local conditions, the nature of the soil, and the tendency of the site to dampness, as well as the permeability of the surface.

The object should be as far as possible to prevent water from reaching the road-bed, and to provide means for immediately removing such as does reach it before the soil becomes saturated and softened.

ART. 10. SURFACE DRAINAGE.

The drainage of the surface of a road is provided for by making the section higher in the middle than at the sides, with ditches or gutters at the edges of the road along which the water is conducted until it may be disposed of through some side channel.

The slope necessary from the middle to the sides of

the road to insure good drainage depends upon the nature of the covering, being less as the road surface is more smooth and less permeable to water. It varies from about 1 in 20 or 1 in 30 for broken stone to 1 in 40 or 1 in 60 for various classes of pavement, and for asphalt sometimes as low as 1 in 80.

The form of section used is commonly either a convex curve, approximately circular, or it is made up of two plane surfaces sloping uniformly from the middle to the sides in each direction, and joined in the middle by a small circular arc. There has been considerable dispute among engineers as to which of these forms is most desirable, although the general preference seems to be given the plane section. It is not usually a matter of special importance, provided the section used is not too flat at the middle for good drainage, and not too steep at the gutters for safety. In places where surface-water must be carried for considerable distances in gutters at the side of the road, and provision must be made for a considerable flow, the gutters may be deepened by increasing the slope of the surface at the sides, or rounding off as much as possible without making the slope too steep for safety.

The road should also have a certain longitudinal slope in order that the water may flow freely in the gutters. This slope should be at least 1 in 200 in most cases in paved streets, and somewhat greater—about 1 in 100 to 1 in 120—on broken-stone or earth roads. Where longitudinal slopes are steep, some provision must be made to prevent the wash of the gutters, and in such places it is specially desirable to take the water from the gutters as frequently as possible, in order to make the gutter flow small. This may often require, where no sewers exist, the laying of a special pipe underground for the purpose.

On country roads the disposal of surface-water is not usually a matter of difficulty, as it can be carried along the road and run into the first convenient cross-channel.

In towns the most satisfactory method of disposing of surface drainage is through a system of storm sewers, the water collected in the gutters being emptied at frequent intervals into the sewers and thus quickly removed from the surface of the street. In the absence of such a system it may often be necessary to lay pipes, connecting with the nearest natural channel, to relieve the gutters. In such cases catch-basins should always be placed at the entrance to the pipe to prevent it getting clogged by the dirt which may be washed in from the gutter. On lines of pipe of considerable length, catch-basins should also be introduced at intervals, to allow the accumulated sediment to settle and be removed.

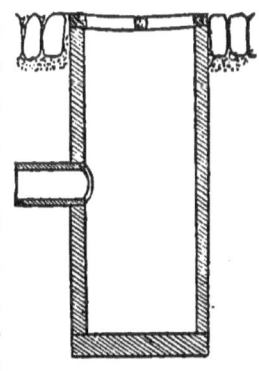

Fig. 3 represents a basin of this kind. It may be formed for small pipes, of a length of pipe set on end with the lower end closed, or where necessary a box built of masonry may be employed.

FIG. 3.

In all cases it is important that the water which falls upon the surface should be gotten rid of as soon as possible, for so long as it remains upon the road it it is an element of danger, both from its tendency to wash the surface, and from its liability to penetrate into the road and thus cause disintegration or settlement. The best method of removing this water in any particular case must be determined by a careful study of local conditions, and its final disposal in the case of the streets of a town is a special problem requiring careful treatment.

Art. 11. Subdrainage.

The drainage of the sub-soil of a road-bed may be directed either to the removal from the road-bed of water that percolates through the road covering, or to the prevention of sub-surface waters from reaching and saturating the road-bed.

The necessity for subdrainage, and the method to be employed in any case, depends upon whether the soil over which the road is being constructed is naturally wet or dry, and whether the road-bed is so situated and formed as to give it natural drainage.

Where artificial subdrainage is necessary the drains should be located, in so far as possible, with a view to cutting off the supply of water before it reaches the road-bed. To accomplish this to the best advantage the local conditions must be observed, the sources of this supply determined, and the nature of the underflow, if any exist, considered.

In many situations, particularly when the site of the road is low and naturally damp in wet weather, it may be advisable to place a longitudinal drain under each side of the road. Such a construction is shown in Fig. 4, which gives a section of a macadamized country road with tile side-drains.

Frequently, as in many cases of a road along a side slope, there is a well-defined flow of sub-surface water from one side to the other, and in such case the water

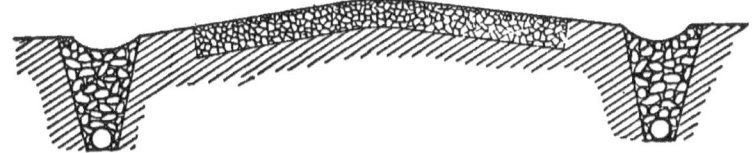

FIG. 4.

may perhaps be intercepted by a single longitudinal drain on the side of the roadway from which the water comes. An example of this is shown in Fig. 5,

FIG. 5.

which represents a macadamized village street with stone curb, gutters, and sidewalks.

In other cases, where the subsoil is of a very retentive nature, or where the natural slope of the land is in the direction of the length of the road, cross-drains leading into a longitudinal side-drain or into side-ditches may be expedient, and sometimes, especially upon narrow country roads, a single longitudinal drain under the middle of the road may give the best results, serving both to remove sub-surface water and that which percolates through the road surface. Fig. 6 shows such a road, representing an ordinary earth road with a tile centre-drain. Fig. 20 also represents a stone centre-drain as sometimes used under a broken-stone road over wet ground.

These longitudinal drains should be arranged to empty as frequently as possible on country roads into the natural drainage-channels, or in towns into sewers arranged to convey the water rapidly away.

In general, systematic underdrainage will not be necessary except in localities where the ground is naturally damp from lack of natural drainage, or where an

Fig. 6.

underflow creates a tendency to wetness in the subsoil. In some localities, however, upon country roads, where an impervious surface is not employed and the soil is one that absorbs and retains water, it may be necessary to provide subdrainage to remove water that passes through the road surface. This is most commonly done by a series of shallow cross-drains or by a single longitudinal one in the middle of the road.

In a town where sewers traverse the streets subsoil drainage is easily arranged for in connection with the sewers. Frequently blind-drains or stone-drains are constructed underneath or alongside the pipes. In other cases good drainage is secured by drains underneath the curb or gutter, which are connected with the sewers.

Art. 12. Kinds of Soil.

The material of which a road-bed is composed is important because it determines to a large extent whether artificial drainage is necessary, and also what method should be adopted for securing drainage.

Soils differ in their power to resist the percolation of water through them, in the rapidity and extent of their absorption of water with which they come in contact, in the extent to which moisture renders them soft and unstable, and in their power of retaining moisture.

A light soil of a sandy nature usually presents little difficulty in the matter of drainage, as, while it is easily penetrated by water, it is not retentive of moisture, which passes freely through it without saturating it unless prevented from escaping.

If the natural drainage, therefore, have a fall away from a road-bed formed of such material it will usually need no artificial drainage, and where subdrains are necessary they may be relied upon to draw the water from the soil to a considerable distance each side of the drain.

A nearly pure sand is more firm and stable, under loads, when quite damp than if dry, although a fine sand saturated by water which is unable to escape may become unstable and treacherous.

Clays usually offer considerable resistance to the passing of water through them, and are very retentive of moisture. As a rule, however, a clay soil does not absorb water readily, and requires that water be held for some time in contact with it in order that it may become saturated, although when saturated it is the most unstable of soils. A clay that when dry will stand in a vertical wall and support a heavy weight when wet may lose all coherence and become a fluid mass. When water comes in contact with a bed of such clay, the outside becomes saturated and semi-fluid before the moisture penetrates into it sufficiently to even moisten it a few inches from the surface.

A clay soil is, therefore, always difficult to drain by removing the water after it has soaked in, or by permitting it to pass through the road-bed to the subdrains beneath. Drainage, in such cases, may often be so arranged as to prevent water from standing against the road and thus prevent it from becoming saturated. As the clay is comparatively non-absorptive, the water which may come upon its surface, if allowed to escape at once, will not penetrate into it, and hence will not cause softening.

A heavy silt formation is sometimes met with which is even more difficult to drain than a true clay. It is nearly as retentive of moisture as a clay, strongly resisting the passage of water through it, but at the same time absorbs water quite freely when in contact with it.

Between the extremes mentioned above there are a great number of varieties of soil which possess to a greater or less extent the characteristics of either or both, and gradually merge the one into the other. In applying a system of drainage in any case, careful attention should always be given to the characteristics of the soil, as determining very largely the treatment to be used.

In pervious, sandy, or gravelly soil drains may often be effective for a distance of 30 or 40 feet through the soil, while with a less pervious retentive clay the drain may not act effectively more than 8 or 10 feet on **each** side.

Art. 13. Types of Drains.

For the purpose of draining the subsoil of the road-bed the drains used may be either open ditches at the sides of the roads or porous covered drains.

Open ditches are sometimes used on country roads. They are usually placed at the extreme edges of the road, and must be deep in order to be effective. Being so far from the travelled portion of the road, they can only act satisfactorily as subdrains where the soil is pervious and easily drained.

In other cases, where side-ditches are employed, covered cross-drains must be introduced to carry the water from the middle of the road to the ditches. Fig. 7 shows a section of a country road drained by side-

FIG. 7.

ditches. Where such ditches are employed the slope of the sides should be made as gradual as possible, at least $1\frac{1}{2}$ or 2 horizontal to 1 vertical, in order to diminish the danger of the washing of the banks, as well as the liability to overturning of a vehicle over the edge.

Covered underdrains are to be preferred to open ones for this use, and are more commonly employed where efficient subdrainage is attempted. These drains must be so arranged that they may be readily penetrated by the water without becoming clogged by earth washing into them. The types of drains most commonly employed for this purpose are known as blind drains, box or stone drains, and tile-drains.

For short lengths, such as transverse drains intended to take the water from the subsoil into the side ditches or drains, *blind drains* may frequently be economically employed. They consist simply (as shown in Fig. 8) of ditches cut into the soil and filled with rounded stones 3 to 6 inches in diameter. Angular stones are not desirable for this purpose, as the object is to leave openings into which the water may penetrate without difficulty, and thus be led away. The top of the stones must be covered over in some way before filling in earth above in order to prevent the earth from washing down and choking the drain. This is sometimes accomplished by using smaller stones at the top, covered by a layer of coarse gravel. A layer of straw or brush, or of sod turned roots upward to retain the earth until it becomes thoroughly compacted, is a common and effective method of protecting these underdrains.

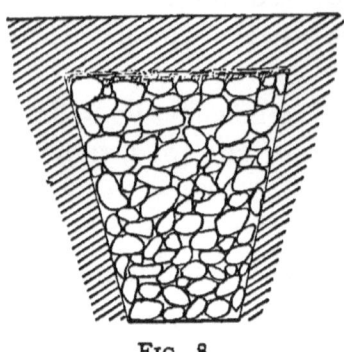

FIG. 8.

Stone-drains are commonly employed where stone is plenty and cheap. They usually consist of rectangular or triangular boxes formed of flat stones or bricks at the bottom of a trench, which is then filled as in a blind-drain so as to give ready access for water.

Figs. 9 and 10 show sections of stone-drains as commonly constructed of rough field-stones. The form given in Fig. 9 is commonly known as a *box drain*.

Tile-drains are probably in general the most convenient and efficient for subdrainage. They are made

DRAINAGE OF ROADS AND STREETS. 33

of round, or U-shaped, unglazed drain-tile, laid, as in the last case, at the bottom of a ditch filled with round stones.

Fig. 11 shows the section of a tile-drain as commonly constructed.

The tiles are usually set end to end in the trench,

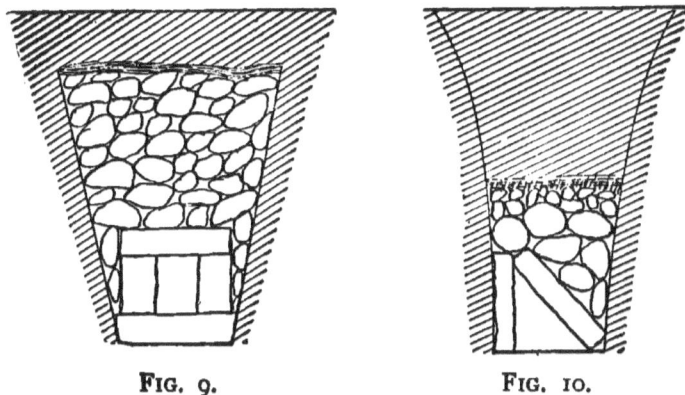

FIG. 9.   FIG. 10.

being held in place by small stones braced underneath.

The joints are thus left open to permit of the free entrance of water. Collars for the joints may be obtained and are sometimes used where thought necessary to keep larger material from washing into and obstructing the tile. These collars are rings of pipe into which the ends of two adjoining sections of the tile may be fitted, and they thus serve also to hold the tile in line.

The filling of the trench above the tile, as in the other drains, should be arranged with a view to maintaining a

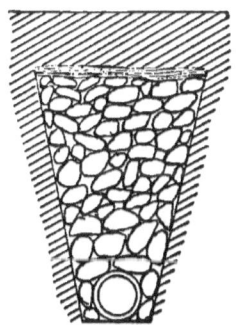

FIG. 11.

porous structure through which water may easily pass. Sometimes flat stones are laid over the tile resting with one edge on the bottom of the trench and meeting above at the middle so as to form an additional protection to keep any earthy matter from choking the entrance to the tiles.

All of these drains should be deep enough to escape freezing.

The mouth of a tile-drain should also be protected in some manner, as the porous tile is apt to be broken and destroyed by frost when saturated. In some cases the tile-drain is made to discharge through a short length of stone-drain, or through a section of salt-glazed sewer-pipe, which will not be injured by freezing.

It is desirable in all cases to protect the mouths of underdrains with a netting of some kind to prevent the entrance of vermin, which may clog the openings with their nests.

The slope of a porous drain may vary from about 1 in 40 to 1 in 100. In case a steeper slope be necessary, a foundation or paving should be placed in the bottom of the trench, which is otherwise liable to be eroded by the current that may be produced.

Art. 14. Culverts.

Culverts are commonly required in road construction for carrying under the road the small streams which may be crossed by the road, or sometimes for carrying the water collected in the gutters or ditches on the upper side of the road to the lower side.

The waterway provided by a culvert must, for

safety, be sufficiently large to pass the maximum flow of water that is likely to occur, while for economy it must be made as small as may be without danger.

The maximum flow of a stream depends upon a number of local conditions, most of which are very difficult of accurate determination. These are: the maximum rate of rainfall; the area drained by the stream and its position; the character of the surface drained; and the nature of the channel.

The maximum rate of rainfall varies in different localities, and differs in the same locality from year to year. It is commonly taken at about an inch an hour. This is sometimes exceeded for a very short time and over a small area, but is usually a safe value for a watershed of any considerable area.

The approximate area of the watershed drained by a stream is readily found, and its form is also important as determining the distance the water must flow in reaching the culvert under consideration, and to some extent regulating the rate at which the water falling upon the area will reach the culvert.

The maximum flow of a stream is also affected by the physical characteristics of the watershed. The permeability of the surface largely determines what portion of the rainfall shall reach the stream; while the slope of the surface, its evenness, and its vegetation have an effect upon the quickness and rate with which the rainfall is received by the stream.

The determination of the maximum flow to be expected in any case from an examination of the locality is therefore possible only as a very rough approximation. A number of formulæ have been proposed for such estimation, the use of which for the case of an ordinary culvert simply amounts to estimating the quantity of water which would fall on the watershed in the heaviest probable rain, and judging as well as possible from local conditions how much of it may arrive at one time at the culvert. In some cases where a more accurate determination is desirable it may be advisable to measure the flow of the stream at high water, and form an idea from such measurement as to what may be expected at a maximum stage.

The amount of water that will pass a culvert in a given time depends upon the form of the section, the smoothness of its interior surface, its slope, and the head under which the water is forced through. A well-constructed culvert may be considered in computing its capacity as a pipe flowing full. Other culverts or bridges must be treated as open channels.

Prof. Talbot gives (Selected Papers C. E. Club, Univ. of Illinois, 1887-8) a formula for the rough determination of area required for waterway, derived from experience:

$$\text{Area waterway in feet} = C \sqrt[4]{(\text{drainage area in acres})^3}.$$

C is a coefficient depending upon local conditions. For rolling agricultural country subject to floods at time of melting snow, and with length of valley 3 or 4 times the width, $C = \frac{1}{3}$. When the valley is longer, decrease C. If not affected by snow and with greater lengths, C may be taken at $\frac{1}{5}$, $\frac{1}{6}$, or even less. For steep side slopes C should be increased.

For most cases in practice the size of waterway required may be determined from the knowledge which usually exists in the vicinity regarding the character of

DRAINAGE OF ROADS AND STREETS. 37

a stream, from the sizes of other openings upon the same stream, or from comparison with other streams of like character and extent in the same locality. Where data of this kind do not exist, careful examination of water-marks on rocks, the presence of drift, etc., may be made to determine the height to which water has previously risen.

For small flow of water box culverts of stone or pipe culverts are commonly employed. Wooden box culverts are also sometimes used, constructed of planks or heavy timbers, but should be avoided in so far as possible on account of their perishable character, and consequent lack of economy.

Pipe culverts are the most efficient in use, and as they can now be constructed quite cheaply in most parts of the country, are coming into very general use. The efficiency of a pipe culvert may frequently be greatly increased by arranging it to discharge under a considerable head at times of unusual flood. This requires that the water shall freely flow away below the outlet, and that the surface of water above the culvert may stand higher than the head of the pipe.

Pipe culverts may be constructed either of salt-glazed vitrified sewer-pipe, or of iron water-pipe. The iron pipe possesses greater strength, and is preferable where a firm foundation is not easily obtained, as it is not so easily broken by any slight settlement.

In laying pipe culverts, they should be placed on a solid bed, and the earth be well tamped about them. It is desirable to have the bottom of the trench excavated to fit the lower part of the pipe, depressions being formed for the sockets. It is necessary in every case that the pipe be firmly and uniformly supported from below, in order that the culvert may not be broken by settlement, which is especially likely to occur in new work.

It is desirable that the joints in the pipe be made water-tight, especially where the culvert is likely to flow full or under pressure, as any water escaping through the joints will tend to cause a wash beneath the pipe and undermine the culvert. Joints are commonly filled with clay. Where strength is needed the use of hydraulic cement mortar is preferable, and sometimes the small end of the pipe is roughened on the outside and the socket on the inside in order that the cement may hold more firmly.

Care should be taken that the culvert have sufficient slope and be so placed that water may not stand in it, in order to prevent injury from freezing, and the top of the culvert pipe should be at least two feet below the road surface to avoid crushing.

The ends of pipe culverts should be set in masonry walls to give protection against the washing of the face of the embankment, hold the ends firmly in place, and prevent the entrance of water into the earth on the outside of the pipe.

These walls to give efficient protection must be of substantial construction, going down to a solid foundation below the bed of the stream. They may be built of rubble masonry, and should be laid up in hydraulic cement mortar. Such construction is represented in Fig. 12. The wall must extend far enough on the side to sustain the earth of the embankment from the waterway, or wing walls may be used extending up stream for this purpose. The waterway should be paved above the culvert far enough to prevent scouring at the base of the wall.

For quite small streams the walls may sometimes be omitted if the face of the embankment about the entrance to the pipe and the waterway for some distance above and below be riprapped. Where it is necessary to economize in the cost of construction, this method is preferable to the use of very light end walls.

On streams too large for a single pipe it is often economical to lay two or three pipes side by side,

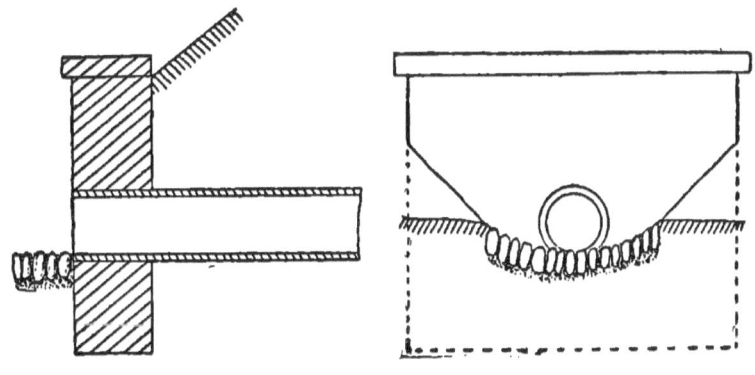

FIG. 12.

rather than to construct an arch or the open way of a bridge. In laying large pipes it is usually advisable to place a broken-stone or concrete foundation under the pipes throughout their lengths to insure uniform support.

Stone Culverts.—Culverts of stone may be either arch culverts or box culverts. Box culverts are usually formed of two side walls and a cover. The side walls consist usually of rubble stonework laid up dry or in mortar, as the case may be. Where the stream to be carried is of small importance, and the capacity of the culvert not greatly taxed, dry walls may give satisfactory results, but when the culvert is likely to flow full at certain times it should be laid up in hydraulic cement mortar, and in any case the greater stability given by the mortar would be well worth the small additional cost. Fig. 13 shows a section of the ordi-

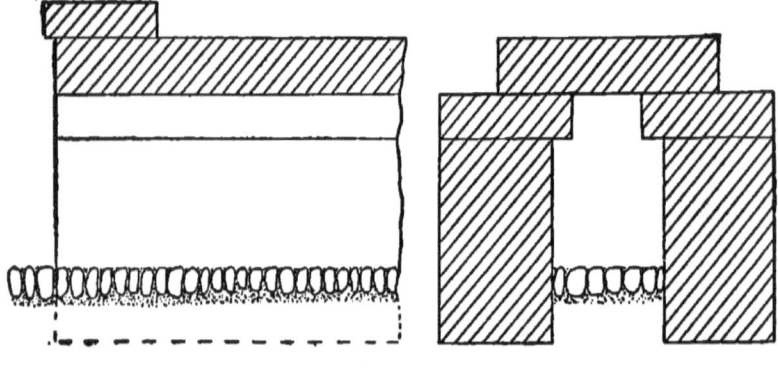

FIG. 13.

nary form of box culvert. The use of head walls and paving the waterway for a short distance is necessary for these as for pipe culverts.

Where suitable stone is available, box culverts are easily constructed and economical. They are commonly used for openings 2 to 4 feet in width and 2 to 5 feet in height. The width that may be used depends upon the available cover stones. Where the allowable width is not sufficient to give the needed area of waterway, a double culvert may sometimes be used to advantage. This consists of two openings with a middle wall to support the covers.

The culvert's opening should always be large enough to admit of a man passing through it for the purpose of cleaning it—at least 18 by 24 inches. The side walls should extend downward below the bottom of the culvert sufficiently to obtain a good foundation, and the thickness required for the side walls usually varies from one half to three fourths the height, depending upon the pressure likely to come against them.

In many cases for small work the side walls, instead of extending downward, rest upon the paving which is extended under them. This gives a somewhat less expensive construction, and is often satisfactory on good ground.

The cover stones may be from $\frac{1}{3}$ to $\frac{1}{2}$ the span in thickness, and should be long enough to have a bearing upon each side wall of at least one half the thickness of the wall.

Arch culverts are used for openings too large to be made of the box form or of pipes. The discussion of arches and also that of bridges in general is outside the proper scope of this book. For ordinary country bridges wooden trusses are most commonly employed, on account of their comparative cheapness. For short span bridges a satisfactory and economical construction, which has recently come extensively into use, consists in placing a number of wrought-iron eyebars across the opening from abutment to abutment at short distances apart. Brick arches are then used to span the spaces between the eyebars, which are tied together with wrought-iron rods, and the roadway is then constructed over the bridge in the same manner as upon the earth road-bed.

Concrete culverts may sometimes be used to advantage where they can be cheaply constructed. They are usually made in oval form, the bottom being first formed by ramming the concrete upon the foundation so as to form a curved channel 4 to 6 inches thick. The upper part is then constructed as an arch upon a centre, which is left until the mortar has set. For the small openings for which these culvert sare employed the thickness of concrete in the arch may be from $\frac{1}{4}$ to $\frac{1}{5}$ of the width of opening. The concrete for such work should be made of hydraulic cement, in the manner employed in constructing the foundations for pavements. (See Art. 47.)

Abutments for small bridges should be laid upon solid foundations, and built of hydraulic cement mortar, the back of the abutment wall being made impervious by coating it with mortar. A common and safe thickness of abutment is to make the thickness $\frac{4}{10}$ of the height. The waterway between abutments should be paved to prevent scouring out the foundations.

Art. 15. Water-breaks.

Upon heavy gradients on country roads, continuous for any considerable distance, water-breaks are commonly placed at frequent intervals to collect the water which flows down the surface of the road and turn it into the gutters or side-ditches. They should only be used on grade steep enough to make them essential, as otherwise they form an obstruction to traffic. They consist of broad shallow ditches, and should be arranged to carry the water from the middle of the road each way to the gutter, thus forming a V with the vertex uphill and at the middle of the road. This arrangement permits teams following the middle of the road to cross the ditch squarely. It is desirable that these cross gutters be paved to prevent washing during heavy showers.

CHAPTER III.

LOCATION OF COUNTRY ROADS.

ART. 16. CONSIDERATIONS GOVERNING LOCATION.

THE determination of a line for a proposed road involves the examination of the country through which the road is to pass with reference to its topographical features, the nature and extent of the traffic that it may develop, and the local interests that may be affected by the position of the road.

The simplest form that this problem can take is that in which two points, as two towns, are to be connected by a road for the purpose of providing for a traffic between them, the nature and amount of which is approximately known. In this case it is only necessary to examine the topography of the intervening country and select the line over which, taking into account the costs of construction and maintenance, the given traffic may be most economically carried.

In most cases in practice, however, the problem does not have this simple character, and in fact location can seldom be determined by considerations of economy alone. The position of the line will be modified by local needs, such as the necessity of providing for the traffic of villages or farms intermediate between the ends of the road, which may often cause deviations from what would be the best line if the interests of the terminal points alone were considered.

Questions of the desirability of various lines for the comfort and convenience of travel, and the pleasure to be derived from the use of the road, dependent upon æsthetic considerations, may also frequently operate to change the line from what would seem proper from a strictly economic point of view.

In thickly settled communities, as in most parts of the United States, the roads are in the main already located, the necessity for the location of new ones does not often arise, and when it does occur is usually mainly determined by the local needs and requirements of traffic.

The *economic considerations* involved in the location of roads are of two kinds: those relating to the accommodation of traffic, and those relating to its economic conduct. The first deals with the necessity of the road to the community, the second with the cost of operating it. The first involves the general question of the advisability of any road, and how it can be placed to give the greatest freedom to the movement of travel. The question is as to the value of the road to the general community and its location to secure the greatest good for the least outlay, without taking into account the details of location which may affect the cost of transportation. The value of the road as developing trade in a town or bringing a farm nearer to market would enter into consideration. The accommodation of traffic requires that a road be located with a view to the convenience of its use by the largest portion of the traffic, as well as with a view of developing traffic.

The position of a road that will best accommodate

traffic is that in which, other things being equal, the mass of traffic need be moved the least distance in reaching its destination; or, in other words, that for which if each ton of freight be multiplied by the distance through which it must be moved the summation of the resulting products will be a minimum. If there be differences in the nature of the routes over which the road may be constructed, they may be considered as equivalent to changes in the relative effective lengths of line for purposes of comparison.

The ordinary problem of location deals mainly with considerations of the second class. It consists for the most part in the relocation of portions of old roads, of making such changes in position when improving a road as may tend to reduce the cost of conducting traffic over it, and render it more convenient and pleasant for the use of travel, or of determining the details of alignment and grade upon a new road which is approximately fixed in position by the purpose of its construction.

The most economical location is that for which the sum of the annual costs of transportation, the annual costs for maintenance, and the interest on the cost of construction is a minimum.

The cost of conducting transportation is affected by the rate of grade of the road, the amount of rise and fall in it, and the length of the road. The rate of grade is important, because it limits the loads that can be hauled over the road, or determines the number of loads that must be made to transport a given weight of freight, as well as fixes a limit to the speed of travel. The amount of rise and fall affects the expenditure of power required to haul a load over the road. The length of the road has an effect upon the amount of work necessary to haul a load over it, the time required for a trip, and the cost of maintaining the road surface; each of which, other conditions being the same, is directly proportional to the length.

The cost of construction depends upon the accuracy with which the line of the road is fitted to the surface of the ground, as determining the amount of earthwork and cost of bridges and culverts; upon the character of the ground over which the road is to be built, which affects the cost of executing the work and determines the necessity for and expense of drainage; and upon the cost of land for right of way. All of these items must be considered in any comparison of the cost of constructing on various routes. Special care should be taken in selecting a line to avoid bad ground, such as swamps, upon which construction may be difficult and expensive. The availability near the line of the road of materials needed for surfacing may also become a matter of importance in the cost of construction, and have an influence in determining location.

The relative importance of the various elements affecting the choice of a line depends upon the nature and amount of the traffic to be provided for, and upon the character of the road surface to be used. Where the traffic is heavy, the importance of reducing the cost of moving it by lessening grades and distance will be greater than where the traffic is light, and the cost of construction may be correspondingly increased for that purpose. If a smooth surface be employed, upon which traction is light, the value of reducing grades will be greater and the value of reducing distance less than with a surface of poorer tractive qualities.

Art. 17. Length of Road.

Changes in the length of a road affect all portions of the traffic in the same manner, and the expenditure of power and loss or gain in time occasioned by them are in general directly proportional to their amounts.

The value of any considerable saving in length may usually be considered as equal to the same percentage of the whole cost of conducting the traffic that the saving in distance is of the whole length. If, therefore, a rough estimate may be made of the annual traffic to be expected upon a given line of road and of the cost of carrying the traffic, this cost divided by the length in miles through which the traffic is moved will give the annual interest upon the sum that may reasonably be expended in shortening the road one mile, or upon the value of a saving of a mile of distance; or dividing by the number of feet of distance will give the value of saving one foot.

It is to be noted, however, that the cost of the work of transportation is not necessarily proportional to the amount of work done, and consequently this method would not be strictly accurate even were the data as to traffic and costs readily obtainable. An estimate of this character at best amounts to only a rough guess, but it may often be of use as an aid to the judgment in deciding upon the value of a proposed improvement involving a considerable change of length in a road.

Where the road is so situated and the saving in distance proposed is such that it would enable teams to make an additional trip per day in the hauling of freight, the difference in cost of transportation is quite tangible and readily estimated; but where the traffic is of a more indefinite nature, or the saving proposed insufficient to admit of additional trips, the value of the difference of length depends upon the value to other work of the small portions of time of men and teams which may be saved by the shorter route—a value which exists, but is difficult to estimate.

There is also a value in the saving of distance due to the advantage to the community of bringing the various points closer together, such as bringing two towns into closer relations or bringing country property nearer to markets. The method of considering the cost as proportional to the work done will therefore probably give a fair idea of the actual economy in any saving in the work of transportation.

The value of reducing distance varies with the character of the road surface. As the cost of transportation is less over a smooth than over a rough surface, on account of the lighter traction, the value of reducing distance is also less on the smooth surface.

The value of saving distance also is greater on a road where the ruling gradients are steep than upon one with light gradients, because of the greater number of loads necessary to move the same traffic.

The cost of maintenance of a road varies with its length, and under similar conditions may be considered, like the costs of transportation, to be directly proportional to the length of road.

The saving in cost of maintenance from decreasing distance must of course be added to that in cost of transportation in order to find the actual value of a change of length.

The value of straightness for a country road is frequently very much overrated. Considerable devia-

LOCATION OF COUNTRY ROADS. 49

tions from the straight line may often be made with but slight increase in length, and there seems to be no good reason for insisting upon absolute straightness. The error is commonly made of sacrificing grade and expense in construction to the idea of straightness without the attainment of any considerable saving in length.

It involves in many cases the injury of the beauty of the road and of the landscape, with no compensating economic advantages.

Art. 18. Rise and Fall.

By the amount of rise and fall is meant the total vertical height through which a load must be lifted in passing in each direction over the road. It is distinct from and independent of the rate of gradient.

The minimum amount of rise and fall is found where the rise is all in one direction and the fall in the other, each being equal to the difference of elevation of the terminal points. Any increase in the rise and fall beyond this amount is represented by the rise encountered in passing from the higher to the lower terminus. It affects the traffic equally in each direction, and requires a certain expenditure of power to lift the traffic through the given rise in each direction.

If the cost of developing the work necessary to overcome rise and fall be the same as that of developing an equal amount of work to overcome distance, the rise and fall may be evaluated in terms of distance, and any change in rise and fall may be considered as though it were a difference in distance and treated as in Art. 17.

Where the rate of grade is less than the angle of repose of the wheels upon the road surface (see Art. 2) the work necessary to overcome rise and fall will be that which will lift the load through a vertical height equal to the amount of rise to be considered. When the rate of grade is greater than the angle of repose, an additional amount of work must be done in applying a resistance to prevent the too rapid descent of the vehicle in going down the grade. The amount of this work in any case equals the work done in lifting the load to a height equal to the difference between the actual rise of the grade in question and the rise of a grade of the same length and a rate equal to the angle of repose. Thus on an ordinary earth road whose resistance to traction where level is 100 pounds per ton, suppose a grade to occur of 8 feet per 100, 1000 feet in length. For the road surface we have $100 \div 2000 = .05$, and the angle of repose is a 5% grade. Then $8\% - 5\% = 3\%$, or the brake-power necessary to secure uniform motion, is the same as would be necessary to haul the load up a 3% grade, and a grade of 3 in 100 for 1000 feet gives 30 feet. The work to be done in holding back the load for the 1000-ft. grade is therefore the same as would lift the load through a vertical height of 30 feet, or the fall of 8 feet per 100 for 1000 feet has the same effect as 30 feet of rise in the same direction, provided brake-power costs the same as animal power.

The value of rise and fall in terms of distance will depend upon the nature of the road surface, as the work necessary to lift a given load to a given height is a constant, while the work done in hauling a load over a given distance will vary with the resistance offered to traction by the surface. Thus, taking the surface as above, the

LOCATION OF COUNTRY ROADS. 51

work of lifting one ton through a rise of 1 foot is 2000 foot-pounds, while with a tractive force of 100 pounds per ton 2000 ÷ 100 = 20 feet, the distance a ton may be moved on the level surface in developing 2000 footpounds of work. Therefore 1 foot of rise or fall may be considered as equivalent to 20 feet of level distance, and the value of reducing the amount of rise and fall may be found from that for reducing distance.

If the road considered were a first-class Macadam road, with resistance of 40 pounds per ton, 1 foot of rise or fall would equal 2000 ÷ 40 = 50 feet of distance.

ART. 19. RATE OF GRADE.

The effect of any change in the ruling gradient upon a road depends to a considerable extent upon what portion of the traffic may be carried in full loads. The lighter portions of the traffic are not so seriously affected by heavy gradients as the heavy portions, although there is an advantage in light gradients for any driving. The rate of speed which may be employed will be less upon the portions of the road having heavy grade, and the time occupied in a trip over the road is therefore affected somewhat by the rate of grade.

The desirability of a road for general driving is also much influenced by the gradients employed, as is that value of the road which has for a basis the effect it may exert upon the attractiveness of the locality. These things all have a certain financial value, which of course it is quite impossible to estimate with any degree of accuracy, but which should be considered in determining the allowable maximum gradient in any case in practice.

For heavy traffic, such as the transfer of goods from one town to another or the marketing of country produce, the limitation of load placed upon the traffic by the gradient is a matter of importance, the effect of which is calculable upon the cost of transportation. If in any case the approximate amount of this heavy traffic which is likely to be carried in full loads be determined, the relative costs of its transportation over two lines of differing gradient, other conditions being similar, will be nearly proportional to the number of loads required to move the traffic over each gradient.

In estimating the value of reducing the rate of grade, it may be considered, as in the case of a reduction of length, that its value to the community is represented by the saving in annual costs of transportation, and that the amount that may reasonably be expended in increased cost of construction to effect a reduction of gradient is the sum upon which this annual saving is the interest.

The length of a road and the amount of rise and fall on it determine the amount of work that must be done in hauling a load over the road. The rate of gradient, on the contrary, does not affect the amount of work necessary to move the traffic, but it limits the work that a horse may do at one trip.

The establishment of a proper rate for the ruling grade of the line is, therefore, usually the most important point in location. In localities where light gradients are easily obtained the problem of location is greatly simplified.

By referring to Article 3 the comparative loads that a horse may draw up different grades will give some idea of the importance of carefully considering the

LOCATION OF COUNTRY ROADS. 53

question of gradient. In nearly all cases in practice there is a considerable latitude within which gradients may be chosen. It is usually a question of heavier gradients as against greater distance and larger first cost for the road. It may be remarked that it is only under exceptional circumstances that it is either necessary or advisable to use a steeper gradient than 5% on the new location of a country road of any importance. Grades steeper than the ruling gradient may sometimes be introduced over short distances without impairing the efficiency of the road, as horses are usually able to exert for a short time a force much greater than they can continuously exert. If the length of grade be quite short, 200 or 300 feet, a horse can about double his ordinary power in passing it.

Where long steep grades must be used, it is desirable to break them by short stretches of lighter gradients to provide resting-places for horses.

Heavy gradients also have the disadvantage of retarding traffic in the direction of falling grade, and, as suggested in Art. 18, of requiring the expenditure of work to hold the load from too rapid descent.

ART. 20. EXAMINATION OF COUNTRY.

For the purpose of obtaining the requisite data upon which to base the location of a road, it is necessary that a careful examination be made of the topographical features of the country through which the line is to pass. The relative elevations of the termini of the line and of intermediate points should be obtained, and the directions and steepnesses of the various natural slopes determined.

If a line were to be located connecting points at long distances from each other, as sometimes occurs in railway location, it would be necessary to study the general configuration of the country, noticing the direction of flow of the streams, and the location and elevations of the various passes in the ridges through which it might be possible to carry the line. Usually, it would be found that the country is composed of a series of valleys, separated by ridges, branching in a systematic manner from the main watercourse of the region, and that the passes in the ridges occur at the head of side streams, and especially where streams flowing into valleys on opposite sides of the ridge have their sources near each other.

In the location of common roads, however, the problem is ordinarily of a less extended nature, and may consist in joining two points lying in the same valley, or in joining points in adjacent valleys by a line passing over a ridge. In these cases it is only necessary to take into account the slope of the valleys in question, the positions and elevations of available passes, and the side slope of the ridges.

The slope of the bed of a valley, in hilly country, usually forms a concave curve, the rate of slope gradually increasing from the lower to the upper end. In a valley of considerable length this increase in the rate of slope may be very gradual or, in short valleys rising to a considerable height, it may be more sudden. The profile $ABCD$ in Fig. 14 shows the slope of a short valley which decreases in slope from about ten feet per hundred at the upper end to about two feet per hundred at the lower end.

When a map of the country to be traversed is available, showing the positions and elevations of the points controlling the location, the work is very much simplified, the reconnaissance may for the most part be limited to a study of the map, and the routes may be sketched upon the map to be tried in the field. If the map at hand is an accurate contour map on a sufficiently large scale, the entire location may be worked out in detail upon the map, leaving only the work of staking out the line to be done upon the ground.

Maps may be obtained, in most parts of this country, upon which the horizontal positions of points may be readily fixed with sufficient accuracy for the purposes of the preliminary examination. Where such maps are not obtainable, the positions of points must be ascertained and a rough map prepared. For this purpose directions may be measured with a pocket compass, and distances estimated or obtained by the use of an odometer or pedometer, as may be most convenient.

Differences of elevation are easily obtained with a fair degree of accuracy by the use of an aneroid barometer, and slopes may be measured with a hand level.

Where the rough means ordinarily employed in the reconnaissance are not sufficiently accurate to determine the controlling points of the lines to be adopted, a more complete examination of the country may often be made by a rapid topographical survey by means of the transit and stadia method.

Whatever means may be adopted for doing the work, the preliminary examination should determine a map showing the approximate positions of the controlling points through which the road must pass, and enable a rough sketch to be made of the slopes of the country through which the line is to be run.

Art. 21. Placing the Line.

After the preliminary examination of the locality is complete and the positions and elevations of the controlling points of the line are known with reference to each other, the line must be selected and run in upon the ground, or, if the reconnaissance is not conclusive as to the position of the best line, it is advisable to run in two or more lines and make a more detailed comparison between them.

The controlling points of a line are those points at which the position of the road is restricted within narrow limits, and is not subject to change. These may be points where the location is governed by the necessity of providing for traffic, or points where the position of the line is restricted by topographical considerations, such as a summit over which the line is to pass a ridge or a favorable location for a bridge.

Where the line is to be located to an uniform gradient, it should be started from the controlling point at the end of the grade, which is usually the summit. It is then laid off along the slope in such manner as to cause it to have continuously the rate of grade decided upon. Taking D (Fig. 14) at the summit of the valley as the controlling point, it is seen that the distance from C to D is sufficient to give a gradient of 10 in 100 by following directly down the valley, and the line with that gradient may be run in that manner.

The maximum gradient from A to C is, however, only 5 in 100, and if thought advisable the same maxi-

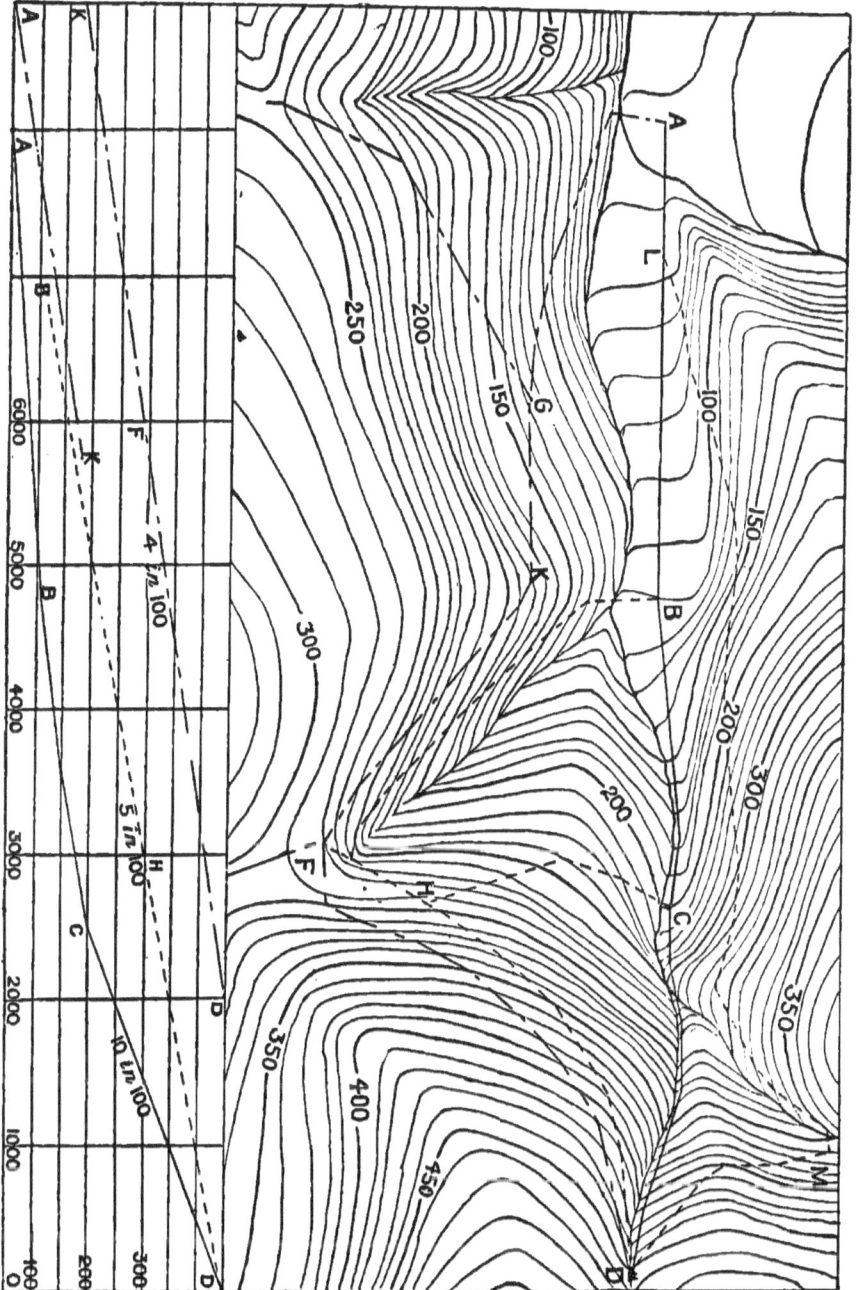

mum gradient may be used between C and D by running the line DHC diagonally down the slope, as shown. This line, having one half the gradient, will have about twice the length of the line CD.

In running this line it is started from the highest point of maximum grade, and points at the surface of the ground are continually selected, in advance of the placing of the line, which are at the proper elevation to permit the grade to pass through them. This may be accomplished by setting off the angle of the gradient upon the vertical circle of the transit, or upon a gradienter, and sighting upon a rod which is moved until the line of sight strikes it at the same height from the ground that the instrument is setting. The points for the line may also be found by running a line of levels ahead of the transit line (a hand level is convenient for this purpose) and pacing distances upon which to reckon the gradient, the distances and elevations being frequently checked upon those of the measured line.

The location of a gradient upon a common road differs from that upon a railroad only in that steeper gradients are used, sharper curves or angles may be employed, and the gradients need not be lessened on ordinary bends or curves. If the line is to make a turn upon the slope as at H, the grade should be flattened at the turn, and a curve of as large radius as possible, without too great expense for grading, be introduced.

In a manner similar to the above a line might be run from D on the other side of the valley, which using a 5% gradient would give the line DML, reaching the bed of the valley at the point L. A lighter

LOCATION OF COUNTRY ROADS. 59

gradient may be obtained from A to D by starting from D and going down by a continuous gradient of 4 in 100 on the line $D\ F\ G\ A$, and greater or less rates of descent may be adopted and lines corresponding to them located, as may be considered advisable.

The centre-line for a final location should be carefully run, and points permanently marked from which it may be relocated when necessary. An accurate line of levels should also be run over the centre-line and a profile drawn, upon which the grades may be established and earthwork estimated.

After placing the centre-line, topography should be taken carefully upon each side of the line for some distance, and a map drawn showing the topography and giving elevations by means of contours. This will serve to show whether the line is placed to the best advantage, and whether any changes are desirable. This is especially necessary over rough ground or where the line is on maximum gradient, as frequently, and perhaps usually, the first line run will be useful only as a preliminary line, which with its accompanying topography will permit a proper location to be made.

ART. 22. COMPARISON OF ROUTES.

In selecting a line for the construction of a road the principles already mentioned in the early part of this chapter should be had in mind. The line must be well designed to accommodate the traffic. It should have as easy grades, short length, and small rise and fall as is consistent with a reasonable cost of construction, in order to give light costs for transportation and for maintenance.

Suppose in the case shown in Fig. 14 that it is desired to connect the village at the point *A* with the point *D* and with the roads leading through the passes at *F* and *I*. Which line it will be the most advantageous to adopt depends upon the relative importance of the traffic to the various points considered.

The shortest, and probably cheapest, line from *A* to *D* would be obtained by following the valley over the line *ABCD*, which line, as shown by the profile, would give a maximum gradient of 10 in 100 between *C* and *D*. The line *FB* joining the first line at *B* would afford communication with the summit at *F* with a maximum gradient of 5 in 100. If the traffic to the point *D* be small and unimportant, so that additional expense in reducing the gradient from *C* to *D* is unadvisable, these lines might prove a satisfactory location.

If, however, *D* be a point of importance and the traffic from *A* to *D* heavy, it will be necessary to adopt some means to reduce the gradient from *C* to *D*. Leaving out of consideration the point *F* and considering *B* and *C* as points of minor importance, it might be advisable to use the line *ALMD* with an uniform 5% gradient from *D* to *L*, and branches to connect with *C* and *B*. This would give a line but little longer than the valley line, with only one half the ruling gradient of that line.

If *C* is not important and can be neglected while *B* and *F* must be considered, the line *ABEHD* has a maximum gradient of 5 in 100, and connects *A* with the points *BF* and *D* with a minimum total length of road (being less than the valley line first considered).

When *B* and *C* must both be considered as of importance as well as F and D, the lines $ABCHE$ and HD will give a ruling gradient of 5 in 100 to both F and D, and passing through B and C with a somewhat longer line than in the last case.

This arrangement would make the length of haul from A to D and F, each longer than by the first line considered; but the gradient to D would be lighter, and the total length of road to be constructed and maintained would be less.

In case the points B and C are both unimportant, and the line through the valley may be neglected, the line $AGFD$ provides a ruling gradient of 4 in 100 from A to both F and D, and connects them with each other, with about the same length as the shortest 5% gradient. When the point I must be taken into account, this line may be connected with I by the line GI having a gradient of 4 in 100. This would give the shortest line of uniform gradient to connect A with the three points IF and D, and possibly a desirable line to construct when the line through the point I is important, even if the valley road from A to B is also necessary.

The lines upon the side slopes are usually more expensive to construct than the valley lines, and the differences of first cost of the various lines must of course be considered. The importance of a difference in expense of construction depends upon the traffic to be hauled over the road and the kind of surface to be used. Where a broken-stone or gravel road is to be constructed at considerable expense, the difference of cost due to a change of location is relatively less important as being a less percentage of the whole cost, while the difference of tractive effort due to grade is of more importance, as being a higher percentage of that upon the level, than would be the case with an ordinary earth road.

As is easily seen from the above the choice of a location for a road, while depending upon principles easily stated, is in reality a matter requiring the use of judgment, and is not readily reducible to a financial comparison stated in money values, because the data concerning the volume of the traffic and the cost of conducting it can be determined only very roughly, and contains many elements of error. For purposes of comparison to aid the judgment, approximate data may often be assumed or determined by a study of the localities affected. In some cases observations may be made of the number of teams of different classes passing certain points within certain times, to give a basis for estimation of the annual volume of traffic. In other cases, the annual hauling traffic, which is usually the most important portion of the traffic in considering location, may be estimated from the known interests of the locality. Thus, if the produce of a certain section of farming country must be hauled over a given road to market, the amount of this produce may be estimated from the acreage, and the relative number of loads upon different grades then determined. The cost per load over the road would then need to be assumed in order to find the annual value of a reduction of grade.

In the same manner, the effect of changes of length and in the amount of rise and fall may be found as indicated in Arts. 17 and 18.

All of these items must be combined to find the relative total costs of transportation for each route. The

LOCATION OF COUNTRY ROADS. 63

cost of construction and of maintenance for each line must then be estimated, and that line is the most advantageous which makes the sum of the annual charges and the interest on the first cost a minimum. Where several lines of traffic are to be considered together as in Fig. 14, the cost of conducting all of the traffic by each system of lines that may be employed must be considered, the entire cost being made a minimum for the system to be adopted.

ART. 23. CHANGING EXISTING LOCATIONS.

The problem that arises oftener than any other in country-road location is that of improving short stretches of road, where, owing to defective location, the grades are unnecessarily heavy, the length unnecessarily great, or the ground over which the road may pass such as to make its maintenance in good condition difficult and expensive. The first of these is the most common defect of ordinary country roads, as shortness of distance has very commonly been obtained by the disregard of the desirability of light gradients, which in very many cases are easily obtainable.

The principles to be observed and methods of procedure in making the new location are exactly the same as in an original location, save that in this case a road already exists, and the question of economy is one of determining whether the advantages to be obtained in lessened cost and transportation and maintenance is sufficient to warrant the expense of attaining new right of way and constructing new road.

In Fig. 15 is given an example that is frequently met in practice, where the existing road *abcd* runs over the point of a hill, with heavy gradient, while a line of very much lighter gradient might be located around the base of the hill through the pass at *e*, giving a greater length of road, but much less rise and fall. The line *bcd* in the figure has a length about 800 feet greater, a rise and fall 70 feet less, and a maximum gradient one half as steep as the line *bcd*. These relations are shown in the profile in Fig. 15.

If the road in question be a common earth road, 1 foot of rise and fall may be taken as equivalent, in the work required to haul a load over it, to 20 feet of distance, and the 70 feet saved by the new location would be equivalent to 1400 feet of distance. Hence, the line *bed* may be considered as having an equivalent length for purposes of traffic $1400 - 800 = 600$ feet shorter than the line *bcd*. In addition to this, loads may be taken over the new line in direction *b* to *d* more than double, and in direction from *d* to *b* triple, in weight those that can be taken by the same power over the old line.

A further improvement of the line may also be possible, if the new line can join the old one at a point lower down than *b*, by running a lighter gradient than 5 in 100 from the point *c*. Thus the line *efa* would give an uniform gradient of 4%, but would require the construction of more new line.

In considering changes of location, it is also necessary to take into account the interests of adjoining owners. Houses and buildings are largely located with reference to the existing position of the roads, and changes in the position of a road may involve injury to such property. The question then becomes largely one of sacrificing the interests of the users of the road,

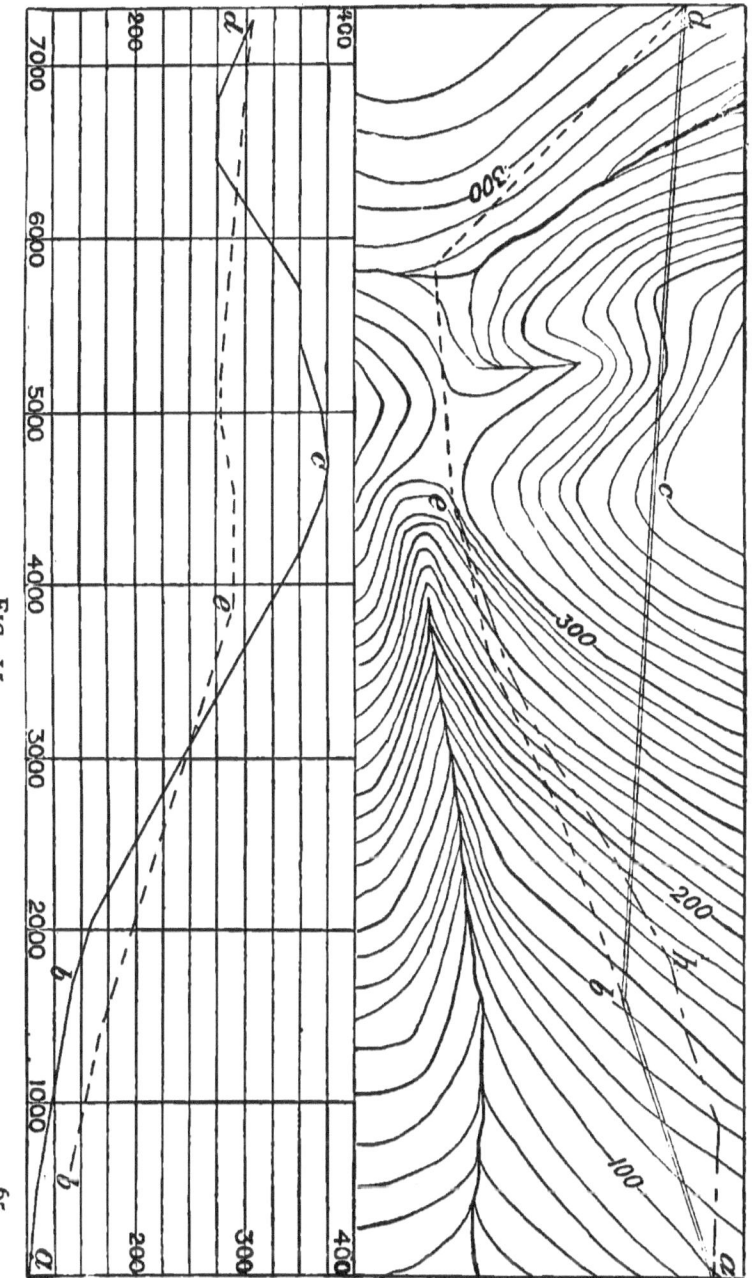

Fig. 15.

or those of the adjoining owners—a question that should be, but commonly is not, decided by considering what will be of most advantage to the general community.

CHAPTER IV.

IMPROVEMENT OF COUNTRY ROADS.

ART. 24. NATURE OF IMPROVEMENTS.

ORDINARY country roads may be classified as earth roads, gravel roads, and broken-stone roads. The larger number of common roads throughout this country belong of necessity to the first class. In a few of the more enterprising communities the more important roads are constructed of gravel or broken stone.

The percentage of roads of the better class is, however, very small and although there has recently been a distinct improvement in this particular, the inability of rural communities to at once raise the funds necessary for the general construction of first-class new roads will cause their increase to be very gradual.

Improvement in country roads may be of several kinds

(1) Changes in location, by which better alignment or better gradients may be obtained, or by which the natural conditions of surface or drainage may be improved. This has been discussed in Chapter III.

(2) Reconstruction of the road-bed, as in regrading steep slopes to give lighter gradients, or in raising the road-bed across low and wet places to provide for drainage.

(3) The construction of artificial drainage where a road is inclined to be wet, as already discussed in Chapter II.

(4) Improvement of the surface, which may consist in reforming the surface of natural earth, or in the construction of an artificial surface or pavement, the latter of which will be discussed in separate chapters.

The problem in common-road improvement is for the most part that of making the most of the roads that exist, rather than reconstructing them with new material. The materials and funds immediately available must be used to secure as much improvement as possible.

Earth roads, under the most favorable circumstances, do not usually attain any high degree of efficiency, and are not economical under any considerable traffic. They are, however, capable of much improvement, and if properly managed need not become, as they frequently do, practically useless during a large portion of the year, although they are always more difficult and expensive to maintain in a good condition than roads of a better and more permanent construction.

Art. 25. Earthwork.

Improvements to the road-bed of an existing country road may have for their object the reduction of gradient upon steep inclinations, by cutting the material from the road-bed and lowering the surface of the road on the upper part of the grade, and filling in correspondingly on the lower part, or they may be intended to provide better drainage by raising the road across low ground.

In the construction of new roads, the formation of the road-bed consists in bringing the surface of the ground to the grade adopted for the road. This grade should be carefully established upon an accurate profile of the line, in such manner as to give as little earthwork as possible, both to render the cost of construction low, and to avoid unnecessarily marring the appearance of the country in vicinity of the road. The most desirable position of the grade line is usually that which make the amounts of cut and fill about equal to each other, especially where room for borrow-pits, or spoil-banks, would be expensive, and it is desirable to make the embankment for the most part of the material taken from the road excavations. On side-hill work, one side of the road is commonly in cut and the other in fill, and where the side slopes are steep, it is usually better to make the road mostly in cut on account of the difficulty of forming stable embankments on steep ground.

Where embankments are to be constructed, the surface of the ground should be cleared of all vegetable matter and soft material before beginning the placing of the earth-filling, in order to give a firm base to the bank and permit it to bond with the earth below. The material of an embankment should be as homogeneous as possible, and all perishable matter should be carefully excluded from it. It should be deposited by beginning at the outside and working toward the middle in such a way as to give a concave section to the top of the bank during construction, which tends to prevent sloughing off along the lines of the joints between the various layers. It is also best to build an embankment a little narrower at bottom and wider at top than it is intended to remain, and afterward trim down the edges to the proper slope.

Earth in an embankment will compact closer than it is found in the natural state. On an average it will shrink about one tenth of its bulk. The allowance to be made for settling in forming an embankment depends upon the method of construction. Where scrapers are used, the earth will usually be well compacted in placing, and no allowance is necessary; with dump carts or wagons the compacting is not so thorough, and a small allowance should be made; while when wheelbarrows are used or the earth is thrown into place with shovels, an allowance of 10 or 12 per cent must be added to the height of the embankment, in order to allow for the final shrinkage. Rock occupies more space in embankment than in excavation, and does not need allowance for shrinkage.

In constructing embankments across wet and unstable ground, it is frequently necessary to form an artificial foundation upon which to place the earth embankment. This may be accomplished in some cases by excavating a little of the soft material and substituting sand or gravel, or in other cases it may be advisable to employ layers of brushwood or fascines as a support for the embankment. Sometimes it may be possible to drain the soft material by deep ditches, so as to render it capable of sustaining the road, and in all cases drainage should be provided in so far as possible to make the embankment more secure.

When embankments are to be formed on sloping ground, the surface of the ground should be stepped off, in order to hold the earth-filling from sliding upon the natural surface at the line of contact between the

IMPROVEMENT OF COUNTRY ROADS. 71

two, until it becomes sufficiently settled for the development of cohesion to cause it to become one solid mass.

In many cases where roads are to be constructed along steep slopes, it is found cheaper to use retaining walls to sustain the road upon the lower side and the earth-cutting on the upper side than to cut long slopes or form high embankments.

Catch-water drains are necessary on the natural surface above the top of all high slopes in cuttings to prevent the surface-water from washing down and destroying the face of the slope.

Where springs are tapped by a cutting, drains must be provided to remove the water without injury to the slope; and where the subsoil may become wet in rainy weather, it may be necessary to provide sub-surface drains along the slope to prevent the earth becoming saturated and sliding down into the roadway.

Slopes, both of excavation and embankment, are greatly improved by being sodded or sown with grass. This aids in the maintenance of the slopes, by rendering them more capable of resisting the abrading action of such water as falls upon them. It also greatly improves their appearance.

The most important principle involved in the formation of a road-bed, which should be always in mind, is that earth in order either to sustain a load or to maintain a slope, must be kept dry, or at least prevented from becoming saturated with water, as both the cohesive and frictional resistances of earth are diminished or destroyed when it becomes wet, and it is also then liable to the disturbing action of frost.

Art. 26. Drainage.

Drainage is especially important upon earth roads, because the material of the road surface is more susceptible to the action of water, and more easily destroyed by it than are the materials used in the construction of the better class of roads. When water is allowed to stand upon the road, the earth is softened and readily penetrated by the wheels. The action of frost is also apt to be more disastrous upon the more permeable surface of the earth road, having an effect to swell and heave the roadway and throw its surface out of shape. It may in fact be said that the whole problem of the improvement and maintenance of ordinary country roads is one of drainage.

In underdraining an earth road on account of the permeability of the surface, provision must be made for carrying off the water which penetrates through the surface, as well as that due to natural wetness of the subsoil. The surface should of course be made of such form and material as to cause the water to flow off without penetrating as far as possible; but in damp weather wheels will mark the surface somewhat, and water held in the ruts so formed will soak into the earth, and unless at once removed below soften it so that the next wheel makes a deeper rut, with the final result of destroying the form of the road as well as its power to sustain the loads that come upon it.

The necessity for the application of artificial subdrainage in any case is determined by local conditions, the character of the soil, and natural drainage. An examination of the line of the road in wet weather, observing whether water stands upon the ground, the direction of flow of surface-water, and whether that which penetrates the ground drains away quickly is usually an efficient aid in forming an opinion as to the necessity for drainage.

The methods employed in draining are considered in Chapter II. Dependence is most commonly placed upon shallow side-ditches, which are seldom of much value except to carry off surface-water; and even when the side-ditches are deep, they can only be efficient for subdrainage when the soil is of a very open, porous nature. In other cases they will not draw the water from the subsoil under the middle of the road, and cross-drains or a centre drain should be provided.

The common neglect of proper drainage is undoubtedly very largely responsible for the general bad condition of country roads.

ART. 27. EARTH-ROAD SURFACE.

The method which should be adopted for the improvement of the surface of an earth road depends upon the nature of the material of which it may be composed. When the material is loose sand, the surface will be more firm if the sand be damp and more unstable in dry weather. In such case a small admixture of clay in the surface layer may give cohesion to the surface when dry, or a layer of clay six or eight inches deep may form a hard and comparatively durable surface, as it is easily drained when upon the sand road-bed.

Clay soils as a rule absorb quite freely the water with which they may be held in contact, and soften when saturated, but are not readily permeable, and hence are not easily drained from below. Used alone they are consequently the least desirable of road materials. When dry, clay may make a very hard and durable surface, and it may give good results as a covering for a road-bed of more pervious material, or it may form a stable road-bed when protected by a surface which does not soften so readily and prevents the surface-water from reaching the clay beneath. In building over clay, sand or gravel may frequently be mixed with the clay to form a surface layer which will be less acted upon by water. When rather coarse sand or small gravel is used for this purpose and a small proportion of clay just sufficient to bind the particles of sand together, a very hard and compact mass is formed, nearly impervious to water and but little acted upon by it. Material of this nature found in a natural state is known as hard-pan, and is very stable and durable. A layer of sand a few inches deep may also sometimes be employed to form a surface over a clay road-bed which will not soften in wet weather, and will afford protection to the clay beneath.

When other material cannot be obtained, clay roads are sometimes improved by burning the clay so as to form a more porous material for use as a surface layer. This method, however, is somewhat expensive, and other materials may usually be employed at less cost.

Soils composed of mixtures of sand and clay or of gravel and clay are usually easier to deal with than clay itself, and commonly form the best natural roads. They vary in character from the light sandy loams to heavy soils partaking very much of the nature of clay. The sandy soils take up water readily and become soft when wet; but they are pervious and easily drained, and they may be compacted into a firm surface in dry weather. The heavier soils take up water readily and become soft when wet, but are less pervious and drained with more difficulty, though much more easily than a clay.

The material of a road surface should always be such as may be compacted to a firm and hard surface. It should not, therefore, be formed of the soft material which may be washed into the gutters. The surface must be formed with a crown at the middle sufficient to shed the water which may fall upon it into the gutters, and prevent water from standing upon the road. The slope necessary to shed the water readily is about 1 in 20, and the most desirable section is usually that composed of two planes of equal inclination, rounded off in the middle and sloping uniformly to the sides, as shown in Fig. 16.

In the construction of an earth-road surface, road-

FIG. 16.

machines or road-scrapers may often be employed to advantage, especially when no grading is to be done other than giving the road the proper crown. The gutters may thus be formed, and the surface shaped up with comparatively little labor.

After the material is in position, the surface should be compacted to the required form by rolling with as heavy a roller as may be available. This is a very important matter in attempting to form a satisfactory earth road, and is almost indispensable to success. If the loose earth be thrown into the middle of the road to be compacted by the wheels of traffic, the action of the wheels will be to cut it, or at least to pack it in a very uneven manner, producing a surface uneven and full of ruts, which will hold water and ultimately cause the destruction of the road. In case, however, the surface be properly rolled, it may usually be made sufficiently firm to hold up the wheels and retain its form under the traffic, and if kept free from ruts until thoroughly compacted will thus be rendered much more capable of resisting the penetration of water and shedding it into the side gutters.

ART. 28. GRAVEL ROADS.

Gravel roads may vary from that in which a thin coating of gravel is used as a wearing-surface upon an earth road to that in which gravel is used as a surface for the heavy Telford construction of a road of the first class. These latter constructions will be treated in Chapter V, under the head of "Broken-stone Roads."

In the improvement of a country road, where the construction of a good Telford or Macadam road cannot be undertaken, a surface of gravel may frequently be used to advantage, giving much better results than could be obtained with the surface of earth. Even a light layer of gravel may frequently prove of very great benefit.

Where the subsoil is of a porous nature and well drained, a layer of three or four inches of gravel, or sometimes even less, well compacted, will constitute a very considerable improvement; especially if, as is usual with these light soils, the nature of the material of the road-bed is particularly unsuitable for the wearing-surface, difficult to compact sufficiently to shed water, and likely to become soft when wet.

Where the road-bed is of clay a deeper layer of gravel, at least 6 inches, is usually required for effective work, as the gravel must be deep enough to prevent the weight of the traffic forcing the surface layer into weak places in the clay beneath, and also to effectually prevent the surface-water from reaching the clay.

Gravel to be used on roads should be sharp and comparatively clean. In order to bind well in the road it should usually have a small admixture of clay. Gravel in which the stones are round or oval, such as is commonly found in the beds of streams, is unfit for the construction of roads; the small stones of which it is composed, having no angular projections, will not bind together, and even when mixed with clay may turn freely, and will be difficult to firmly bed in position. Pit gravel is usually more sharp, but is frequently found mixed with considerable earth, which, as well as the larger stones should be removed by screening before using the gravel. Screens of $1\frac{3}{4}$ inch and $\frac{3}{4}$ inch openings may be employed for this purpose—that material only which passes the larger and is rejected by the smaller being used in the work.

In the construction of a road with gravel surface the road-bed should first be brought to the proper grade, with a form of cross-section the same as that to be given the finished road. The gravel is then placed upon it and rolled to a surface, or left to be compacted by the traffic. It is always advantageous when possible to compact the road by rolling. The road-bed should be well rolled before placing the gravel, and the gravel surface afterward. A smooth hard surface may thus be produced, upon which the wheels of loaded vehicles may roll without producing any visible impression.

Where the compacting of the road is left to the traffic constant watchfulness is necessary to prevent unequal wear and the formation of ruts.

ART. 29. MAINTENANCE OF COUNTRY ROADS.

The maintenance of a country road in good condition is a matter requiring constant care and watchfulness. Any small breaks in the surface must be immediately repaired, and ruts filled and smoothed before they become serious.

The work required to keep a road in repair depends upon the nature of the surface and the efficiency of the drainage. A well-constructed road of good material will be much easier and less expensive to keep in repair than one in which the surface is not firm enough to resist the cutting action of the traffic, or which has a surface compound of material readily softened by the action of water which may fall upon it.

Earth roads under the most favorable conditions are expensive to maintain, and especially so under the common system of repairing once or twice a year, or at long intervals. This system is not only costly in the work required, which usually amounts to a practical reconstruction of the road each time repairs are undertaken; but it is ineffectual in that the road for the larger portion of the time is out of repair and in bad condition, even if the work of construction has been well done, which is not usually the case where this method obtains.

The only way to keep an earth road in good condition is by the employment of men whose business it shall be to continually watch the road, and make such small repairs as may be necessary from time to time. The small washes that may occur during heavy storms, ruts formed by wagons travelling in the same track, or in passing over soft spots when the road is wet, or any small breaks in the surface of the road, should be at once attended to and carefully filled with new material.

Where small repairs are needed over a considerable area of the road the use of the road-machine is usually advantageous, as giving an easy method of smoothing up the surface. The use of a roller is also nearly always of value, both to assist in smoothing the surface to the proper form, and to give compactness to it. By the occasional use of these machines through the dry seasons a road may be kept crowning and hard, so that most of the rainfall will be quickly shed off into the side gutters without injury to the road.

When there are long-continued rains, or when the ice and snow of winter are melting in the spring, an earth-road surface will necessarily be more or less softened and cut by passing vehicles; and at such times a road of this character cannot be maintained in the same condition as in dry weather, or in the condition which would be possible with a less permeable surface, but if at the beginning of the wet period it be in proper form and if the subdrainage be efficient, the injury to the road as well as the duration of the bad condition, will be reduced to a minimum. As soon as possible after such a wet time, the roads should be gone over with the scraper and put into proper form, and then rolled down hard. It is advantageous to have this done before the ground becomes thoroughly dry and hard, as it will work more freely, and may be compacted much closer by the roller than afterward.

In repairing a road where the gutter is filled with soft material which must be removed to afford a free channel for the surface-water, this soft material should not be scraped upon the middle of the road, as it will not form a good wearing-surface. Where, however, a road is in fairly good condition, and merely needs a little smoothing up, it is desirable to work from the gutter, scraping the material lightly toward the middle until the proper crown is obtained.

The difficulty and cost of maintaining a road will of course vary with the nature of the traffic that passes over it. A road for light driving will be much easier to keep in repair than one used by heavy loads, and as the amount of heavy traffic becomes greater the economy of the earth-road surface is lessened, and the desirability of the substitution of a more durable wearing-surface increases.

The width of the wheel-tires upon which the loads are carried is also important in its effect upon the cost of keeping a road in repair. Narrow tires cut and rut the surface of a road, while those of sufficient width act as rollers to compact the material. For the best results the tires should be as wide as possible, and the front and rear wheels of a wagon should not run in the same track. The lighter tractive effort required for wide tires on compressible road surfaces has been referred to in Art. 2.

ART. 30. WIDTH OF COUNTRY ROADS.

The width of the roadway upon country roads should be only sufficient to provide space for the easy conduct of the traffic. For roads of ordinary traffic this requires only that there shall be room for teams moving in opposite directions to freely pass each other. An available width of 16 feet is ample for this purpose, and 14 feet is often sufficient. Too great width in the roadway causes an unnecessary increase in the cost of constructing and maintaining the road. Where the road-surface is of earth it will be much easier to drain it if it be narrow than if it be wide. If deep side-ditches be depended on for subdrainage, the nearer they are together the more effectively will they drain the subsoil under the middle of the road. Side ditches must, however, be far enough apart so that a berm may be left on each side between the travelled part of the road and the ditches. Thus in Fig. 7, p. 31, if the macadamized portion represents the travelled part of the road, the berm between that portion and the ditches could be sown with grass and show the line of the road as a guide to travel.

When covered drains are used for subdrainage the gutters at the side may be made shallow and placed next the travelled part of the road, giving much less surface to maintain and greater efficiency to the drainage than in a wider road. Such sections are shown for earth roads in Figs. 6 and 16. Fig. 4 shows a similar construction with side-drains under the gutter and a broken-stone surface. Fig. 17 shows the ordinary form of a country road with broken-stone surface. On important roads the paved portion is commonly 16 or 18 feet in width, but on roads of lesser importance it may be less, and under light traffic a width of 10 feet may be sufficient—teams, when necessary, turning out upon the sod to pass. Where a less pervious covering is employed, as with gravel or broken stone, width will not have the same tendency to render drainage ineffectual as in the case of an earth road, because comparatively little water will pass through the road-surface to the subsoil. The cost of main-

FIG. 17.

tenance may not, therefore, be so materially affected by the width, although the cost of construction, and hence the length of road, that may be built for a give sum will be directly dependent upon it.

While the improved portion of the road should be as small as is consistent with the proper discharge of the duty required of it, the available right of way need not be so restricted, but should be laid out wide enough to permit of the widening of the used portion when necessary, and allow room at the sides for pedestrians, with a grass border and line of trees. When trees are planted along the roadway they should not be placed so as to form a dense shade over any portion of the travelled road, although a moderate shade is not a disadvantage, and care should be used that they are not near enough to a covered drain to permit the roots to grow into the drain and choke it.

ART. 31. ECONOMIC VALUE OF ROAD IMPROVEMENT.

The value of a road improvement to a community and the amount of money that may reasonably and profitably be expended in the construction and maintenance of common roads is a subject the discussion of which leads different persons to widely different conclusions, depending upon the point of view and the data assumed.

The economic principles involved in a choice of location have already been discussed in Chapter III; and the general value of any other improvement, in so far as it relates to the economic conduct of the traffic, may be considered in the same manner. Any improvement, either in position or surface, that has the effect of increasing the loads that may be taken over a road by a given power lessens the number of loads necessary to carry the traffic, and effects a saving in time and labor of men and teams, which may reasonably be considered to have the same money value as the time used in the work.

On ordinary country roads in dry weather, the amount of load that can be hauled is usually determined rather by the grades than by the nature of the surface. Unless the gradients are very light the amount of load that can be carried on a broken-stone surface does not differ greatly from what may be taken on a dry and hard earth road. In improving a road by substituting a hard surface for a surface of earth the gradients and location should therefore always be carefully studied, with a view to deriving the full practical benefit from the hard surface in the light traction that it may require with easy ruling gradients.

It is in wet and muddy weather that improved surfaces have their chief advantage over earth roads, and the main object of introducing hard and impermeable surfaces is to eliminate the period when ordinary earth roads are apt to be muddy and practically useless for the purposes of transportation, and to substitute a road that may be used at any season. Systematic drainage has a similar object. To a farming community the economic advantage of a road uniformly good at all seasons is greater than might appear at first glance. It may in many instances amount practically to a saving equal to nearly the entire cost of hauling, by permitting the work to be done at times when other work is impossible, thus making men and teams available for other duty in good weather. The ability to use a road at any season is also of advantage in the independence of weather that will make it possible to take advantage of the condition of the markets in the disposal of produce or purchase of supplies.

The nature of the roads has likewise an important effect upon the social life of the people in a rural district, and has much to do with the desirability of a locality as a place of residence. These items all have a real importance, which, while difficult to estimate in money values, show at once in the fact that prices of country property are largely affected by them.

The nature of the country roads affect the towns to which the country is tributary as well as the country itself. They directly affect trade in seasons of bad weather, both in regulating the demand for supplies for country consumption and in controlling the supply of produce which is available for market; indirectly also the prosperity of a rural district means that of its trade centre.

All of these points must be considered in any attempt to arrive at any proper conception of the advantages of a proposed improvement. In any particular case the local interests will determine the relative importance of the various elements, and a careful analysis of the trade that does pass over the road and that would pass over it under different conditions will enable a judgment to be formed as to the value of improvements.

The money spent in road improvement is to be considered as an investment, which will return annual interest to the community in reduced costs of transportation and greater freedom of traffic and travel.

ART. 32. SYSTEMS OF ROAD MANAGEMENT.

Several different systems for managing the work of constructing and repairing country roads have been proposed or are in use in various places. These systems differ in the placing of the control of the roads and in the methods adopted for providing funds.

The control of the roads under the various systems may be vested in the national government, in the various State governments, in county or parish organizations or in townships or districts. In regard to the location of control and responsibility, it may be remarked that there are two points to be kept in view.

1st. In order that the work may be economically conducted, the section of country included under one control should be sufficient to warrant the permanent employment of a man, or corps of men, whose business it shall be to continually look after the roads, study their needs, and systematically conduct their improvement. It should admit of the ownership and use of labor-saving machinery for the economical execution of the work, but should not be large enough to require an elaborate and complicated organization.

2d. The control of road work should be so arranged that, as nearly as possible, all of the interests directly affected by the condition of any road shall have a voice in its management and contribute to its support.

Common roads are essentially local in their character and are not usually employed as lines of continuous transportation over any considerable distance. They are not, therefore, of State or national importance as lines of communication, although as factors in the general welfare of the people they must, of course, like all other such factors, be of general interest and concern to both State and nation.

The nation, and in most cases in this country the State, is too large an unit to assume direct control of road work. In general, the interests over so large an area are so varied, and the requirements so different, as to prevent a harmonious and successful organization of such work with a probability of economical administration. In some cases, however, such control might be wise and proper, and the recognition of the importance of road improvement to the general welfare of the State, through the payment by the State of a portion of the cost of permanent improvements, has in some instances proved a powerful stimulus to local action.

The control of road management by towns and small districts is nearly always inefficient because the organization is too small to support a proper management or provide the necessary appliances for economic work. Under this system the man in charge of the roads is usually engaged in other work, he is not a road engineer, and can, and is expected to, give but little attention to the road work. This system of control is also usually unfair, except in case of roads intended for the accommodation of the local district only. For instance, a road passing through a town may be a thoroughfare for the towns upon each side. The principal traffic may be this through-trade to points beyond the limits of the town in which the road is situated. The cost of keeping up this road is largely due to outside traffic, and the intermediate town should not be required to bear all the expense of maintenance. On the other hand, the interests of the towns whose trade passes over the road are largely affected by its nature, and the people of these towns should be permitted a voice in determining the character of the road. Most of the more important roads of every vicinity pass thus through several towns, and the system of improvement by small districts works injustice both ways—upon those who are obliged to keep a road for the use of others and upon those who are obliged to use a road they cannot cause to be kept in proper condition.

County management seems more successful in this country than any other, as a county, or two counties combined if necessary, is usually strong enough to secure intelligent management and homogeneous enough to have common interests.

The proper management of the common roads in any community requires both experience and intelligence. A man to be efficient in such work must be able to make or modify location where necessary, judge of the value of various materials for purposes of construction, determine the necessity for and means to be adopted for drainage, and possess the executive ability to manage men and control scattered work. The work in each locality is a problem by itself, to be solved by careful study of the requirements of the community, taking into account the local natural conditions and available materials and means.

CHAPTER V.

BROKEN-STONE ROADS.

Art. 33. Definition.

BROKEN-STONE roads consist essentially of a mass of angular fragments of rock deposited, usually in layers, upon the road-bed or a foundation prepared for it, and then consolidated to a smooth and uniform surface by means of a roller or by the action of the traffic which passes over it.

There are two commonly recognized systems of constructing broken-stone roads, differing in the nature of the foundation employed, and known respectively by the names of the men who first introduced them into English practice as Telford roads and Macadam roads.

Each of these systems has been greatly modified in use since the time of its founder, and each name is now used to cover a general class of constructions differing very materially within itself as applied in the practice of different engineers. Each of the systems also has its earnest advocates, who contend for its exclusive use, and numerous controversies have been the result, at the conclusion of which each party is " of the same opinion still." The view taken by different road-builders in this matter, it may be remarked, appears to be the result usually of the local necessities of the vicinities in which they work, and of the skill with which the different systems have been applied in work which has come under their observations. In road-building, as in any other class of engineering works, no rigid rules can be laid down for universal application; each road must be designed for the place it is to occupy and the work it is to do.

In some parts of this country natural gravel is substituted for broken stone in the construction of these roads, the methods of construction being the same as in using broken stone.

Art. 34. Macadam Roads.

Macadam roads as commonly constructed consist of two or more layers of broken stone, each layer being rolled to a firm bearing before placing the next. The broken stone is usually placed directly upon the earth road-bed.

In constructing a macadamized roadway, the road-bed is first brought to the proper grade in the usual manner, and rolled to a uniform surface. The surface of the road-bed is either flat or raised at the middle to the same section as is to be given the finished road-surface. The inclined form is usually employed, and seems preferable on account of affording better drainage in case any water finds its way through the surface layer.

On village streets where curb and sidewalks are employed, this section of the road-bed may extend to the curbing (as shown in Fig. 5), but on country roads a bench of earth should be left at the side between the broken stone and the gutter in order to confine the broken stone while it is being compacted, and prevent the spread of the surface materials. The form of the road-bed before placing the stone would then be as shown in Fig. 18, where the completed road is to be of the form given in Figs. 4 and 7. Where the road-bed is in embankment, it is common to construct the earth embankment to the height of the finished surface, and afterwards excavate the material necessary to admit of

Fig. 18.

placing the surface layers. The embankment should be allowed to settle and become thoroughly compacted before the broken stone is placed upon it, and it is desirable with new embankments that they be used for a short time by the traffic upon the earth surface before finishing the road; where, however, the material is well compacted in construction and can be thoroughly rolled this is not necessary.

In constructing the road-bed its proper drainage must be considered, and where necessary to prevent its becoming wet under the broken stone some means should be adopted to artificially drain it.

Upon the completion of the road-bed, a layer of broken stone, usually from 3 to 5 inches in thickness, is placed upon it and thoroughly rolled. Upon this a second layer is placed and likewise rolled to an uniform surface. Sometimes a third layer is added, or in case of a very thin road it may consist of a single layer, the number of layers depending upon the thickness of the road. When no roller is used, the stone is usually spread on the surface of the road-bed to the full thickness desired for the road, and left to the action of the traffic.

The upper layer constitutes the wearing surface of the road, and upon this it is usually necessary to place a thin layer of finer material called *binding material*, which may consist of rock chips, sand, small gravel, or sometimes loam, and is washed and rolled into the interstices of the rock, with the object of forming a compact and impervious surface. Binding material is in like manner often added to the lower layers of the road, although this has not been common practice. The object should be to fill the voids in the rock as completely as possible, serving to make the road one solid mass, to bind the rock more firmly together, and to prevent the percolation of water through the surface.

When a road is to be constructed over a heavy soil not easily drained and apt to be wet and soft, a foundation consisting of a thin layer of sand or gravel may frequently be employed to advantage. This foundation layer will serve to prevent the stones of the lower stratum of macadam from being forced downward into the soft material of the road-bed, or the material of the road-bed from forcing upward into the interstices of the broken stone. This foundation may consist of a layer of sand or gravel from 2 to 5 inches thick, and should be well compacted by rolling before the placing of the broken stone.

ART. 35. TELFORD FOUNDATIONS.

The distinguishing feature of a Telford road is its paved foundation. It consists essentially of a pavement of stone blocks set upon the road-bed and covered with one or more layers of broken stone.

In forming a Telford road the road-bed is constructed in the same manner as for macadam, being made either level or crowned. A pavement is then placed upon the road-bed from 5 to 8 inches thick, depending upon the thickness to be given the road material, the general practice being to make the pavement about two thirds of the total thickness of the road. The stones used for the pavement may vary from 2 to 4 inches in thickness and 8 to 12 inches in length; they are set upon their widest edges and with their greatest lengths across the road. The irregularities of the upper part of the pavement are then broken off with a hammer, and all the interstices filled with stone chips and wedged with a light hammer so as to form a completed pavement of about the thickness required.

Upon this pavement the layers of broken stone are placed, and the road-surface completed in the same manner as for a Macadam road.

The practice of Telford was to grade the road-bed flat, and then construct his pavement deeper in the middle than at the sides, using for a roadway 16 feet wide stones about 8 inches deep at the middle and 5 inches at the sides. This practice is still followed by some engineers, but it is now more common and usually considered preferable to make the surface, of the road-bed parallel to the finished surface and the pavement of uniform thickness. Fig. 19 shows a section of Telford road as now commonly constructed.

Some engineers in constructing Telford foundations do not roll the road-bed, but simply bring it to grade, and then lay the pavement by bedding the stones in the surface of the road-bed sufficiently to bring their tops to the proper height, in which case it is unnecessary to trim off the tops with the hammer as in the common practice.

An objection sometimes urged against the Telford foundation is that if the foundation be of hard stone it will cause the material above to be crushed by the loads which come upon it, and that greater durability in the wear of the road metal will be obtained by having a

FIG. 19.

more yielding foundation. The durability of the Telford road has, however, been established by long-continued usage. There is no apparent reason why a firm foundation should cause greater wear at the surface, and the materials below the surface are never crushed in the destruction of any broken-stone road.

The relative value of the two systems must always be determined by the local conditions under which a road is to be constructed and the necessity for such a foundation in the particular case.

ART. 36. CHOICE OF FOUNDATION.

The proper foundation to be used for a broken-stone road depends upon the nature and condition of the road-bed upon which it is to be constructed and the nature of the traffic to pass over it. If a firm, well-compacted, and thoroughly drained road-bed may be obtained, of material which will not readily soften under the action of moisture, there will usually be no need for a special foundation, but the first layer of the macadam may be placed directly upon the surface of the road-bed. If, however, the road-bed is of a material retentive of moisture, not thoroughly drained, and likely to become soft in wet weather, and the broken stone be laid immediately in contact with it, the stones of the lower layer of macadam may be gradually worked down by the weight of the traffic into the soft earth, and the soil at the same time work up into the voids in the stone, causing a gradual disintegration of the road. It may thus also become retentive of moisture and subject to the disrupting action of frost. In this case some foundation must be provided which is capable of resisting the penetrating action of the soft material of the road-bed, and of distributing the load over it. This may be the Telford foundation as described in Art. 35, the sand or gravel foundation mentioned in Art. 34, or the Telford foundation upon a layer of sand or gravel, depending upon the extent of the difficulty to be met.

It is not intended in the above to imply that the use of a foundation of this character should take the place of proper drainage. The advisability of artificial drainage should always be carefully considered, and where the road is threatened by water which may be removed by the construction of drains they should be used, but frequently thorough drainage is difficult or doubtful, and it is desirable to adopt heavy construction such as the Telford foundation gives.

In some instances it may be possible, by drains under the road and substituting porous material immediately under the broken stone, to use light macadam superstructure and do away with the necessity for the Telford pavement in difficult soils. Thus in Fig. 20 a construction is shown applicable to wet and unstable soils, the space over the centre-drain and under the middle of the macadam being filled with large rounded stones, which secure drainage and form a stable bed for the broken stone.

It is commonly claimed by the advocates of the Macadam system of construction that on any well-drained and well-compacted road-bed there will be no tendency on the part of the stone to work down or of

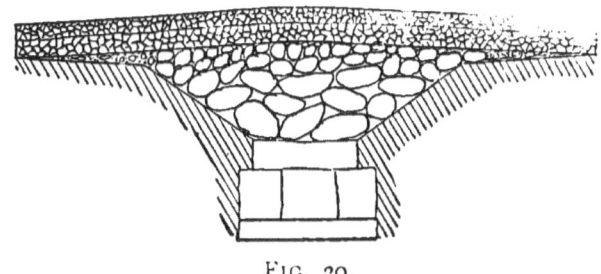

FIG. 20.

the soil to work up, and hence that the Telford foundation is an unnecessary expense. The difficulty of procuring a perfectly stable and reliable road-bed in many localities is, however, very generally recognized, and Telford pavements are largely used.

It would undoubtedly be an advantage in the construction of any broken-stone road, either Macadam or Telford, to have a layer of sand or gravel between the road-bed and the pavement, both as assisting drainage and as providing against unequal settling of the foundation of the road. The application of it, however, must in any case depend upon its cost and its apparent necessity. In cases where drainage is difficult and the soils inclined to be damp and soft, it may frequently prove the simplest solution of the problem.

Concrete foundations are often recommended for broken-stone roads, and would undoubtedly be very beneficial in most cases, but usually where so expensive a foundation may be employed a better surface might advantageously be used than broken stone. It may sometimes occur, however, that, in places where the foundation is difficult to maintain, a light bed of concrete may prove of great benefit, as forming a firm and impervious base to rest upon damp and unstable soils.

ART. 37. MATERIALS EMPLOYED.

A stone to be durable in the surface of a road should be as hard and tough as possible. The qualities of toughness and resistance to abrasion are of more importance than hardness and resistance to crushing. A stone may be hard and brittle, and quickly pound to pieces in a road surface, or it may have a high crushing strength and grind away quickly under abrasion, as is the case with some varieties of sandstone. If, however, it be too soft, it may crush under the loads coming upon it, and thus lack in durability.

A stone for a road-surface must also resist well the disintegrating influences of the atmosphere. It should be as little absorptive of moisture as possible in order that it may not be liable to injury from the action of frost. Many limestones are objectionable on this account.

Basalt and syenite are, in general, the best materials for this purpose. The harder limestones in some localities make a good and durable surface. Soft limestones crush under the action of the wheels, and soon become dust and mud. Sandstones as a general thing are not fit for this use.

The material of a road surface should also be uniform in quality; otherwise the wear of the surface will not be even, and depressions will appear where the softer material has been placed.

As the under parts of the road are not subject to the wear of the traffic, and have only the weight of the loads to sustain, it is evidently not important that the foundation or lower layers be of so hard or tough a material as the surface; and hence it is frequently possible, by using an inferior stone for that portion of the work, to greatly reduce the cost of construction.

A judgment as to the value of any given stone for road use can ordinarily be formed from what may be known of its behavior under other uses to which it may have been subjected, or from its appearance where it has long been exposed to the weather, together with such physical tests as may be necessary to determine if it possesses the special properties desired.

Tests may be made of the power of absorption by drying a sample, weighing it, then placing it in water and reweighing occasionally until it ceases to gain in weight. The absorption may then be expressed as a percentage of the dry weight. The resistance to abrasion may be found either by grinding a sample upon a polishing-disk, or by rattling blocks of the material in an abrasion-cylinder with pieces of iron, and noting the loss of weight in each case, and comparing such losses with those of a stone of known value under like conditions. These examinations with tests of the crushing strength of the material will enable an approximate idea to be formed as to the probable wearing qualities of the stone.

The best test of enduring properties will always be a knowledge of the stone as used for any purpose where it undergoes exposure.

The selection of a stone for road construction will of course always depend largely upon what is to be obtained in the locality of the work. The importance of a thoroughly good material in the road surface is, however, so great in its effect upon the durability and cost of repairs of the road that it may frequently be found economical, on roads subjected to a considerable traffic, to bring a good material a considerable distance rather than to use an inferior one from the immediate vicinity. It may also be suggested in this connection that in many instances railway transportation over a considerable distance may be small compared with wagon transportation over a short distance, and the importation of good material may add but slightly to the aggregate cost of the work.

The size to which stone should be broken for road material depends to some extent upon the nature of the material. The harder and tougher it is the smaller the pieces may be without danger of crushing or shattering under the loads and shocks received in the road surface, and the smaller also they will need to be in order to be thoroughly compacted in the road.

It is a general custom to use larger stones in the bottom courses of a road than at the top. A rule very commonly given is that the stones for the lower layers should be at least 2 inches in their greatest diameter, and not more than 3 inches, and that for the surface layer the stones shall not be greater than 2 inches in greatest dimension.

If of very hard rock the surface layer may have $1\frac{1}{2}$ inches as an upper limit of size.

The size of the rock in the lower layers does not seem of so great importance as that for the surface layers, as it is not directly subject to the weight or the abrading action of the concentrated wheel-loads, and it is probable that in some cases unnecessary expense is incurred in following the refinements of rigid specifications in this particular.

There is a difference of opinion also among roadbuilders as to the advisability of using stone of uniform size. Some insist quite strenuously upon this point, and carefully screen their stone with the object of getting it as uniform as possible; while others declare that the variation of size is an advantage, and even that the stone should not be screened after coming from the crusher, except to remove the stone above the limiting size and when necessary to get rid of foreign matter in case it should contain clay or earth.

Uniformity of size probably makes the wear more even, but the presence of smaller fragments facilitate the binding together of the material. The best practice seems to favor the exclusion of the fine material from the stone, without insisting on too great uniformity in size (stones being allowed probably from $\frac{3}{4}$ inch to $1\frac{1}{2}$ or 2 inches in dimension), and then adding small material after the placing of the stone upon the road to assist binding.

This eliminates the danger of having portions of the road composed entirely of fine material.

Road stone may be broken by hand or machine.

Hand-broken stone is usually preferred as cleaner and better in shape for compacting in the road. In England hand-breaking is largely practised, and it is frequently asserted that machine-breaking injures the stone by crushing it in the jaws of the machine with the effect of decreasing its durability in the road surface.

In American practice machine-breaking is almost exclusively used. It gives satisfactory results, both as to binding and durability, and has the advantage of greatly lessening the cost of construction.

Gravel is frequently used for roads constructed in the same manner as with broken stone, both with and without the Telford foundation. The requirements of a good gravel for this purpose are the same as for a good stone. The stones of the gravel should be sharp and angular, and must possess the qualities of hardness and toughness. Water-worn material is therefore objectionable, as it will not compact without the use of large amounts of soft binding material. In many places a hard flint gravel occurs which is not inferior to the best broken stone.

Gravel not fit for surface material has sometimes been used to advantage under a surface layer of hard rock, and in some cases a surface of flint gravel has been used upon bottom layers of soft rock.

Gravel should be screened to remove the larger stones and the fine material, and then treated in the same manner as broken stone.

Blast-furnace slag may also in some localities be used as a road material with good results. Slag varies greatly in its properties, some being porous and brittle, other hard and tough. The ordinary slag is usually a good material for foundation and lower layers of the road; and where a good, tough slag can be obtained it may also be used for the surface. In some places slag is toughened for such use by being cooled slowly under a cover of ashes or cinders and afterward broken like rock.

Ashes or cinders may also sometimes be employed as a foundation for a thin-broken stone road. They serve to secure good drainage and become very compact. In many cases a considerable improvement might be made in village streets by substituting a layer of ashes for an earth-road surface, and later a stone surface could be applied directly to the road so formed.

ART. 38. BINDING MATERIAL.

It was the practice of McAdam to require that all the stone used upon his roads should be as nearly as possible of a uniform size, and that no foreign substance be mixed with it. In more recent practice it has been found advantageous to use a certain amount of finer material to fill the interstices between the stones, and thus aid in the compacting of the road as well as render it less pervious.

There is considerable difference of opinion upon this point among road-builders. A few still advocate the system of McAdam. Others place a thin layer of binding material upon the surface of the road and work it into the surface voids, while still others distribute the binding material through the entire mass of stone composing the road.

It is agreed that an impervious surface cannot be formed of blocks of hard broken stone without the addition of some small material to fill the voids. It has also been found that when the rock is hard, such as is needed for good wear in a road surface, it will compact with difficulty, and that a certain amount of binding material is necessary in order that the road may be brought to a surface.

The stone forming the body of the road should be placed and partially compacted before the addition of the small material, which may then be worked into the spaces between them.

The office of the binding material is to hold the stones in place and form a bearing for them, as well as to prevent the passage of water between them. It has no duty to perform in sustaining the loads. This is the objection to having the binding material mixed with the stones in advance, as would be the case when unscreened stone is used. A portion of the road stones would be replaced by small material instead of having this material only in such voids as necessarily exist between the stones.

The quantity of binding to be used is that which will be barely sufficient to fill all the voids in the larger material. It has been contended that the lower portion of the road should be porous in order to facilitate the escape of any water that may find its way through the surface, but the tendency of the best modern practice is in the direction of filling all the voids as nearly as possible, thus making the entire road practically one solid body and it is now commonly agreed that the surface of a properly constructed broken-stone road is very nearly impervious to water.

The voids in loose broken stone comprise about 40 to 50 per cent of the volume. In the stone when compacted in the road the voids are somewhat reduced, probably ranging from 30 to 40 per cent of the volume. The voids may be approximately determined in any case by filling a measure with the stone, shaken down as closely as possible, and then measuring the quantity of sand that can be added in the same manner.

Binding material may consist of the screenings from the broken stone used in the road, of sand or small gravel, or of loam. To produce the best results in wear the material used for binding should be of equal hardness with the road stone. Sand, or sand mixed with screenings, often gives satisfactory results, and is more easily compacted than screenings alone. With the use of loam it is much easier to compact the road to a satisfactory surface than with harder material, especially if the roller used be a light one. Loam has been used in some instances with satisfactory results, but the wisdom of its use is questioned by most authorities.

ART. 39. COMPACTING THE ROAD.

The materials may be compacted in a road either by placing them in position and allowing the traffic to pass over them or by rolling with a steam or horse roller.

The first method by itself is seldom practised when it is possible to avoid it. It is hard upon the traffic, takes a long time to reduce the road to a compact condition, and a smooth surface is with difficulty produced. Where heavy horse-rollers are employed they are clumsy and inconvenient to handle, and the work of rolling is slow as compared with the steam-roller. In many instances, however, good results are obtained with them. They are not so expensive in first cost as steam-rollers, and have not the disadvantage of frightening horses.

Some road-builders prefer light-horse rollers of two or three tons weight, using them either by themselves or in connection with the travel. In the latter case the roller simply smooths off the surface as the traffic does the compacting. (See paper by James Owens, in Transactions American Society of Civil Engineers, Dec. 1892.) These rollers are said to work satisfactorily where soft binding material is used, although longer time is necessary than with heavier ones.

Horse-rollers are usually arranged so that the direction of motion may be reversed without turning the roller itself around, and also so that the weight may be changed by placing additional weight inside the roller or removing it.

Steam rollers weighing from 8 to 15 tons are most commonly employed for compacting the road materials. They have the advantage of forcing the materials at once into a firm and compact mass and producing a smooth surface for the immediate use of travel. They admit also of the use of hard materials for binding.

In constructing a road with the use of a steam-roller, the road-stone is first put on to the required thickness and the roller passed over it to settle the stones into place and reduce the voids as much as possible. The binding material, representing a volume about equal to the voids in the stone, is then added, sprinkled, and rolled until the small material is washed and forced into the interstices, giving a smooth, hard surface. This is repeated for each layer of stone, or in some cases the small material is only applied to the top layer.

A thin coating of the binding material is then spread upon the surface and the road thrown open for travel.

ART. 40. THICKNESS OF ROAD-COVERING.

The thickness nessary for a road-covering depends upon the amount of the traffic it is to bear and upon the nature of the foundation afforded by the road-bed. Under a heavy traffic it is advisable to make the road covering heavier than might be allowable for lighter traffic, in order to provide for wear and lessen cost of renewals.

When the road-bed is firm, well drained, and not likely to soften at a wet season, it will always afford a firm bearing, upon which the covering may rest. The loads coming upon the road are then simply transmitted through the covering to the road-bed beneath, and there is no tendency on the part of the loads to break through the covering other than by direct crushing of its material. If, however, the road-bed may become soft in wet weather, it will then lose its power to firmly sustain the covering at all points, and the covering must possess sufficient strength to bridge over places where it is not supported from beneath, or a load coming upon it may break through by bending it downward at such point. The thickness of road covering, therefore, must be greater where the road-bed is less perfect.

The intensity of freezing that may be expected also has an influence upon the necessary thickness of the road-covering. The effect of frost upon the road will depend in large measure upon the condition of the road-bed, and thus make the thickness depend in still greater measure upon its nature. Freezing will not injure a dry road-bed, but if it be damp and have but a thin covering the road is likely to blow or be thrown up by the action of frost.

For roads on considerable grades the thickness of the road-covering is often reduced below what is used on flat ones, because of the better drainage afforded by the slopes. It is to be remarked, however, that if the slopes are very steep the wear of the surface becomes so great, due to the horses' efforts to obtain foothold and to the washing of surface-waters during rains, that the thickness of the coating should be increased.

Macadam roads are commonly made from 4 to 12 inches thick, and Telford roads from 8 to 12 inches, of which 5 to 8 inches may be foundation pavement.

A covering 8 to 10 inches thick is usually considered ample for nearly any case of a country road, unless laid upon bad foundation. In case of a slope of more than 3 or 4 degrees this may perhaps be reduced to 6 inches, or with an especially good road-bed and good drainage it may even be made 4 inches.

A thin road to be effective must have its interstices well filled with binding material and be thoroughly compacted by rolling. It will then present no voids to be filled by the soil pressing upward from below, and at the same time it will be practically impervious and prevent surface-water from reaching the road-bed, thus keeping the material in good condition to sustain the loads. The 4-inch roads of Bridgeport, Conn., which are often cited as examples of successful work, are constructed in this manner of exceptionally good material. In other cases where thin roads have proved failures the trouble may often be traced to dampness in the subsoil or to lack of thorough construction.

In many cases the problem to decide, in determining the thickness of a covering, is whether to use heavy construction or thorough drainage. It is easier to get good results with thick road-coverings, and they are in general safer to use; but skilful adaptation of less material may often save expense in construction with good results. The peculiar conditions of each case must decide what is best for that case.

ART. 41. CROSS-SECTION.

The side-slopes necessary to enable the water which falls upon a broken-stone surface to drain off readily to the gutters is about 1 in 30; hence a crown should be given the road that will permit that slope. In construction the crown should be made an inch or two inches, depending upon the width and thickness of the road, higher than it is intended to remain, in order to allow for settlement and wear under the traffic.

The arrangement of a cross-section has been referred to in Art. 30, and sections are shown in Figs. 4, 5, 7, 17, and 20. In some cases on country roads only the middle portion of the road is surfaced with broken stone, and a roadway of earth is left on each side, between the broken stone and the ditches, upon which teams may drive in dry weather. Such a roadway would commonly be preferred by teams when the earth is dry and hard, but it renders the road more expensive to maintain.

ART. 42. MAINTENANCE OF BROKEN-STONE ROADS.

To maintain a broken-stone road in good condition it is necessary first of all that it be frequently cleaned of mud and dust, and that the gutters and surface drains be kept open to insure the prompt discharge of all water that may come upon the surface of the road.

The best method of making repairs that may become necessary to the road-surface depends upon the character of the material composing the surface and the weight of the traffic passing over it.

If the road metal be of soft material which wears easily, it will require constant supervision and small repairs whenever a rut or depression may appear. Material of this kind binds readily with new material that may be added, and may in this manner frequently be kept in good condition without great difficulty, while if not attended to at once when wear begins to show it will very rapidly increase, to the great detriment of the road. In making repairs by this method, the material is commonly placed a little at a time and compacted by the traffic. The material used for this purpose should be the same as that of the road-surface, and not fine material which would soon reduce to powder under the loads which come upon it. By careful attention to minute repairs in this manner a surface may be kept in good condition until it wears so thin as to require renewal.

In case the road be of harder material that will not so readily combine when a thin coating is added, the repairs may not be so frequent, as the surface will not wear so rapidly and immediate attention is not so important. It is usually more satisfactory in this case to make more extensive repairs at one time, as a larger quantity of material added at once may be more readily compacted to a uniform surface, the repairs taking the form of an additional layer upon the road.

Where the material of the road-surface is very hard and durable, a well-constructed road may wear quite evenly and require very little, if anything, in the way of ordinary small repairs until worn out. It is now usually considered the best practice to leave such a road to itself until it wears very thin, and then renew it by an entirely new layer of broken stone placed in the same manner as in original construction, on top of the worn surface, and without in any way disturbing that surface.

If a thin layer only of material is to be added at one time, in order that it may unite firmly with the upper layer of the road it is usually necessary to break the bond of the surface material before placing the new layer, either by picking it up by hand or, if a steam roller is in use, by means of short spikes in its surface. Care should be taken in doing this, however, that only the surface layer be loosened, and that the solidity of the body of the road be not disturbed, as might be the case if the spikes are too long.

CHAPTER VI.

FOUNDATIONS FOR PAVEMENTS.

ART. 43. PREPARATION OF ROAD-BED.

IN forming a road-bed upon which to place a pavement, the earth should be brought at subgrade to the form of the finished road-surface, leaving room for the superstructure of uniform thickness to be placed upon it. Thorough drainage must of course be carefully attended to when necessary. This has been already discussed in Chapter II.

The road-bed after being brought to the proper grade should be thoroughly compacted by rolling before placing the pavement. Sometimes in the use of a heavy roller, when the material is of a light nature, it is shoved forward in a wave before the roller and refuses to become compacted, in which case a thin layer of gravel or small stone placed upon the surface of earth before rolling may have the effect of consolidating the road-bed under the roller to a hard surface.

If the material over which the pavement is to be constructed is a retentive clay which would become soft when wet, it is sometimes desirable to excavate the clay to a depth of 6 inches or 1 foot below the grade of the road-bed and fill in with sand or some other porous and non-retentive material and consolidate this to form the road-bed. This would seem unnecessary in case a sand or gravel foundation is to be used, and its necessity in any other case would depend upon the likelihood of the road-bed becoming wet, either through natural wetness of the soil or in consequence of the use of a pavement with open joints or of pervious material. If the clay substructure can be kept dry it will sustain the loads which may be carried by the road and need not be replaced.

In constructing a road-bed to bear a pavement the same principles would be involved as in the earthwork of a common road which has been discussed in Art. 25.

Art. 44. Purpose of Foundation.

The chief object of the foundation or base of a pavement is to distribute the concentrated loads which come upon the surface of the road over a greater area of the usually softer and weaker road-bed, in order that these loads may not produce indentations in the surface.

In a foundation composed of independent blocks extending through its thickness, as in the case of a stone-block pavement in which the blocks rest directly upon the road-bed or upon a thin layer of sand, the load which comes upon the top of any block will be distributed over the area covered by the base of the block.

Where the foundation is composed of small independent particles, like sand or loose rounded gravel, with no cohesion through the mass, the pressure is distributed over the base of a cone whose vertex is in the point of application of the load, and the inclination of whose elements depends upon the friction of the particles of the material upon each other. In this case the area over which the load is distributed varies directly as the square of the thickness of the foundation. Sand, it is to be observed, has also the property, when confined as in a foundation, on account of its incompressible nature, of adjusting itself to a uniform pressure and resisting the deformation of the road-bed.

If the small pieces composing the foundation are cemented together, or held as in masses of angular fragments by the interlocking of the angles, the foundation may act more or less as a whole, causing a distribution of the load over a considerable area, the extent of which will depend upon the resistance of the mass to bending.

The bases most commonly employed for pavements are sand, broken stone, and concrete. Foundations of brick and wood are also frequently employed for pavements of the same materials.

Art. 45. Sand Foundations.

Sand foundations as sometimes used under block pavements consist simply of a bed of sand from 6 to 12 inches deep, spread over the road-bed and thoroughly compacted by rolling.

To obtain the best results the sand should be placed in layers of 3 or 4 inches depth and each layer be rolled before the addition of the next. This insures the equal consolidation of the entire foundation. Under a heavy roller the sand may usually be compressed to $\frac{3}{4}$ or even $\frac{2}{3}$ of its original volume.

Well-constructed sand foundations are often very efficient for quite severe service.

Where, as is very common, pavements are set on a bed of loose sand, subsequent settlement of the base is likely to take place, even if the surface of the pavement be well rammed during construction.

Block pavements are also frequently set upon a thin layer of loose sand. This, however, can hardly be considered a sand foundation, as the sand only acts as a cushion to protect the earth of the road-bed from direct contact with the blocks. It also at the same time may facilitate drainage.

ART. 46. BASES OF GRAVEL AND BROKEN STONE.

Foundations of gravel and broken stone are constructed in much the same manner as those of sand. Small gravel will act in much the same manner as sand. In general, however, as a base for a pavement composed of independent blocks these larger materials are inferior to one either of sand or concrete.

These foundations may also be constructed by the use of binding material sufficient to thoroughly consolidate the mass, after the manner of broken-stone roads, and when carefully made are quite efficient in use.

In Liverpool a base has been largely used under stone-block pavements, which consists of broken stone grouted with coal pitch and creosote oil, then covered with quarry-chippings and thoroughly rolled. This is said to give satisfactory results on streets of moderate traffic, and is cheaper than hydraulic concrete.

Art. 47. Concrete Bases.

The best base for general use under pavements is without doubt that formed of good hydraulic cement concrete. A bed of concrete made of good hydraulic cement, well rammed and allowed to set and harden, becomes a practically monolithic structure, nearly impervious to water and possessing a high degree of strength against crushing.

The concrete is formed of a mixture of cement, sand, and broken stone or gravel. The proportions vary for different work and with the character of the materials. With good Portland cement, the most common proportions for ordinary work are about one part cement, three parts sand, and five to seven parts broken stone. With the various natural cements the proportions vary from the above to one part cement, two parts sand, and three of broken stone.

In preparing the concrete the cement and sand should first be thoroughly mixed dry, then sufficient water added to reduce the mass when well worked to a sufficiently plastic condition to be coherent. It should not be made soft or semi-fluid. The amount of water necessary to give the proper consistency should be first determined, and then this quantity added each time to the mixed sand and cement. The mortar should then be reduced to a plastic condition by working it, and not by the addition of more water. The water should never be applied to the mortar from a hose.

When the mortar has been well mixed it is added to the broken stone, which should first be dampened by sprinkling sufficiently to wet the surfaces and wash the dust from them. The mass is then to be mixed until the mortar and stone are uniformly distributed through it. This is commonly done by turning the whole mass over several times with shovels.

The concrete, when well mixed, is placed in position and rammed. The more thorough the ramming the better.

The foundation, after completion, is allowed to remain several days before the pavement is placed upon it, five or six days are usually sufficient, in order that the mortar may become entirely set. During setting the concrete should be protected from the drying action of the sun and wind, and should be kept damp to prevent the formation of drying cracks.

The sand used for mortar should be clean and sharp. The broken stone should be limited to $2\frac{1}{2}$ or 3 inches in largest dimension, and in some case is limited to 2 inches.

Small gravel is sometimes used in place of sand for the mortar to form concrete, although the smaller material is preferable. Gravel is also sometimes mixed with the broken stone in order that a less quantity of mortar may be necessary to fill the voids in the stone, thus making an impervious but weaker concrete.

In some cases foundations are prepared by placing the broken stone and the mortar upon the road-bed in alternate layers, and mixing by ramming them into position. A layer of stone is first placed and wet, then a thin layer of mortar; a second layer of stone is added and rammed into the mortar, after which other layers are placed and rammed in like manner. This method has been followed for the best class of work in Liverpool, and is reported as giving good results.

Frequently in laying asphalt pavements, a concrete is used for the base, which is formed by mixing asphaltic or coal tar paving cement with the broken stone. This is known as a *bituminous base* and is similar to that mentioned at the end of Art. 46. It is commonly constructed by placing the broken stone upon the road-bed to a depth of about 4 or 5 inches and rolling it to a firm and even bearing, as in the construction of a broken stone road, after which a coating of coal-tar cement is given to it, about one gallon of the cement being required for a square yard of the base.

Art. 48. Brick Foundations.

Foundations of brick have frequently been used under brick pavements. The pavement in such cases consists of two layers of brick, with sand between, and is known as *double-layer pavement*. These foundations are usually formed by placing upon the road-bed a layer of sand or gravel 3 or 4 inches thick, which is rolled thoroughly to a uniform surface, and then receives a layer of brick, commonly laid flat and with the greatest dimension lengthwise of the street. These bricks are laid as closely as possible with broken joints. The joints are filled with sand carefully swept in, and the bricks are rammed to a firm bearing.

Upon this course of brick is placed a cushion layer of sand, and then the surface layer. The bricks of the lower layer may be of a cheaper grade than the surface-paving brick, as they are not required to resist the attrition of travel.

ART. 49. SAND AND PLANK FOUNDATIONS.

Under many wood pavements, and sometimes under brick surfaces, foundations formed of sand and planks are used. These foundations differ somewhat in construction in various localities, but are essentially a bed of sand or gravel, upon which is placed a layer of tarred boards which support the surface layer.

It is common to use a layer of sand 3 or 4 inches thick, which is compacted by rolling, after which the boards are laid lengthwise of the street close together, so as to form a floor upon which the blocks may be set. With a brick surface a cushion coat of sand is used under the surface layer.

Sometimes two layers of one-inch tarred boards are employed, the lower being laid crosswise of the street and the upper lengthwise of it. In other cases, the boards of a single thickness are nailed to scantling laid across the street and bedded in the sand. The boards must in all cases press evenly upon the layer of sand that covers the road-bed.

ART. 50. DEPTH OF FOUNDATION.

The thickness required for the foundation of a pavement depends upon the nature of the soil upon which it is to rest, and upon the extent and weight of the travel to which it is to be subjected.

On a well-drained road-bed of good material, a depth of 6 inches of concrete is usually sufficient, even under heavy loads. 6 or 8 inches of well-compacted sand or gravel will likewise usually suffice.

When, however, the road-bed is of a retentive material and likely to become wet and soft, the foundation should possess sufficient strength not to be broken through at points where the supporting power of the road-bed may be destroyed by water. It must also be able to resist the action of frost upon the soil below. In such cases 12 inches of concrete may be necessary, or 12 to 16 inches of sand or gravel may be desirable.

Under light traffic with good conditions, a less depth may be sometimes used, 4 inches of concrete is frequently employed to save expense, although 6 is the more common depth.

It is always important that the foundation be sufficient. The yielding of the base of the pavement means its destruction.

If a firm and durable foundation be employed, the surface may be renewed when necessary or changed from one material to another without disturbing the base, but if the base be weak the surface will be destroyed.

The saving of expense should be at the top rather than at the bottom of a pavement.

It may be here observed that no definite prescription for any pavement, either as to choice of foundation or as to methods of construction can fit all cases. What is most successful in one case is quite inapplicable in another. The blind following of particular rules by those not conversant with the principles upon which they are based has been the cause of many failures. Judgment must always be used in weighing the local conditions of the problem in hand.

CHAPTER VII.

BRICK PAVEMENTS.

ART. 51. PAVING-BRICK.

THE requisites for a good paving-brick are that it shall be hard, tough, and impervious, as well as capable of enduring against the disintegrating influences of the weather.

The bricks in most common use are made from fire-clay of an inferior quality, or from an indurated clay or shale of somewhat similar composition.

These clays consist essentially of silicate of alumina, with usually small percentages of lime, magnesia, iron, potash, soda, and sometimes other elements. The range of composition for clays in common use is approximately as follows:

Silica................	60 to 75	per cent.
Alumina	10 to 25	" "
Iron oxide............	3 to 8	" "
Lime	0 to 4	" "
Magnesia.............	0 to 3	" "
Potash...............	0.5 to 3	" "
Soda.................	0 to 2	" "

In a few cases the quantity of lime is greater varying from 8 to 12 per cent.

When the clay is very nearly pure silicate of alumina, it is capable of withstanding a high degree of heat without fusing, and is known as fire-clay. As the percentages of other ingredients increase it becomes more fusible. The lime, magnesia, potash, and soda act as fluxing agents, and the readiness with which the clay can be melted depends upon the relative quantities of refractory and fluxing materials that it may contain.

Silica in excess tends to make the brick weak and brittle, while too great quantity of alumina causes the brick to crack and warp in the shrinking which occurs during burning. The proper adjustment of the relations between these elements is necessary to good results.

The quantity of lime in the clay is an important matter, as the presence of lime in an uncombined state in the brick may be productive of disintegration when the brick is exposed to the weather. A large percentage of lime in a clay is therefore to be regarded with suspicion, although not necessarily as cause for condemnation, as its effect depends upon the state of combination of the various ingredients of the brick. Magnesia probably acts in much the same manner as lime. Potash and soda are considered desirable elements in quantities to properly flux the clay in burning.

The fineness of a clay is also a matter of importance, both because a fine clay will fuse at a lower temperature than a coarse one, and because fineness is necessary to the production of even and close grained brick, and therefore conduces to make them tough and impervious.

To produce a good paving-brick, a clay is required which will vitrify at a high heat. A very refractory clay will make a porous brick, while if it melts at too low a temperature it cannot be burned sufficiently to become hard and tough.

The methods of manufacturing paving-brick vary in different localities according to the character of the material to be worked. They are quite similar to those in use for common brick, only more thoroughly executed.

The clay is commonly reduced to a fine powder, tempered with water and passed through a machine that moulds the bricks, which are then dried and afterward burned. Repressed bricks are those which are compressed in a mould after coming from the brick machine and before drying.

The process of burning occupies usually from 10 to 15 days.

The heat is at first slowly applied to expel the water, then raised to a high temperature for several days, after which the bricks are very slowly cooled.

There is considerable difference of opinion among engineers and manufacturers as to the exact amount of burning necessary. It is usually stated that the brick should be burned to the point of vitrification, but not completely vitrified. The burning must be thoroughly done to produce a strong and impervious brick, but there is undoubtedly a point beyond which, for some brick, further burning causes brittleness. Very gradual cooling is also necessary in order to toughen the brick.

Smoothness and uniformity of texture in a paving brick is an important consideration as affecting its resistance both to crushing and to abrasion. The broken surface of the brick should present a uniform appearance both in texture and in color.

All of the bricks used in the same pavement should also be of the same degree of hardness and toughness in order that the pavement may wear evenly, and to this end careful inspection should always be given to the bricks proposed for use, and all of those which are defective, soft from imperfect burning, brittle from overburning or quick cooling, or cracked, should be rejected.

The usual and most convenient size for the paving bricks is about the same as that of building bricks. These dimensions, however, vary considerably in practice, and scarcely any two manufacturers make them exactly of the same size. The ordinary dimensions are from $7\frac{1}{2}$ to 9 inches long, 2 to $2\frac{3}{4}$ inches wide, and $3\frac{1}{2}$ to $4\frac{1}{4}$ inches deep, requiring from 50 to 75 bricks per square yard when set on edge. The corners are sometimes rounded off by a curve of $\frac{1}{4}$ to $\frac{1}{2}$ inch radius, or bevelled off, which is not an advantage on ordinary work where close joints are desirable. There seems to be no advantage to be gained in making larger bricks, as has been proposed. It is difficult to burn a larger brick, a better foothold is given by the pavement with joints close together, and if a firm foundation be secured the 4-inch depth is ample, while the numerous joints introduce no element of weakness.

ART. 52. TESTS FOR PAVING BRICK.

To determine the probable durability of brick designed for use in paving, mechanical tests may be applied which will show the relative rank of different samples in their most important characteristics. The tests ordinarily proposed or used for this purpose are those of crushing strength, transverse strength, abrasion and impact, absorption, and specific gravity.

For the *crushing tests* it is common to use 2-inch cubes, sawed from the brick, and brought to a surface by grinding, without cutting with chisels or hammering the specimens, in order to prevent any injury to the material which might reduce its power of resistance. A sheet of soft paper or a thin layer of plaster of Paris is sometimes used between the ends of the specimen and the compression blocks of the testing-machine to equalize the pressure over the surface of the cube.

The result of a compressive test of stone or brick depends very largely upon how it is made, and the results of tests are only properly comparable with others made in the same manner and with equal care. The use of plaster beds as suggested above, it is thought, conduces greatly to regularity of result in the work of different men, as it tends to reduce the effect of differences in the accuracy of dressing the surfaces of contact. The size of the test-piece is also important, the strength usually increasing as the size increases. Small pieces, $1\frac{1}{2}$ or 2 inch cubes, are usually employed, because of the large force necessary to crush a whole brick, although where machinery exists capable of doing it the larger tests entail much less work in preparing specimens.

It is to be observed that the actual crushing strength of a brick is not a matter of special importance in so far as any danger of the crushing of the material in the pavement is concerned, as no stress can there come upon it under ordinary circumstances which would endanger even a very weak specimen from direct crushing. It is thought, however, that to some extent the value of the brick is indicated by its resistance to crushing, coupled, of course, with a proper examination of its other necessary attributes. A brick which possesses a high crushing strength is not necessarily a good paving brick, as it may at the same time be brittle or of such composition as to easily disintegrate under the action of the weather; but one that yields to a low crushing strength is usually weak in wearing qualities and not fit for the purpose. A good paving brick, in the form of a 2-inch cube, will usually show a resistance to crushing of not less than 10,000 pounds per square inch. Much higher values are sometimes used in specifications, but their advantage is at least doubtful.

The transverse strength is tested upon a whole brick by supporting it on edge upon knife-edges 6 inches apart, and bringing the load by a third knife-edge upon the middle of the brick. Soft paper, cloth, or leather may be interposed between the knife-edges and the brick to prevent the abrasion of the brick at the points of contact. This test is easier to conduct satisfactorily, and probably gives, in general, a more reliable indication of the value of the material than the crushing strength. It calls into play not only the compressive but the tensile strength of the brick. In the conduct of the test care is, however, quite as essential as in the crushing test. It is especially important that the brick shall have a perfectly even bearing upon the supports before the application of the load, in order that it may not be subjected to a twist under the load.

The modulus of rupture for the material may be deduced by the ordinary formula $R = \dfrac{3}{2}\dfrac{Wl}{bd^2}$, in which

BRICK PAVEMENTS. 125

R is the modulus of rupture in pounds per square inch, W is the breaking load at the centre in pounds, l is the length, b the breadth, and d the depth of the specimen, all in inches.

The modulus of rupture of paving bricks of good quality ranges from 1500 to 3000 pounds per square inch.

The fracture of a tough and homogeneous specimen under a transverse load should be a clean break through the middle of the brick, and a close observation of the breaks may frequently be of considerable assistance in forming an idea of these qualities, although they may not be directly represented by the load required to break the specimen. The shattering of the brick in breaking, or irregular breaks extending from the point of application of the load to one of the points of support, are apt to indicate brittleness of the material.

Tests for *abrasion* and *impact* have been conducted in various ways, and there is no standard method by which the value of the material to resist these forces may be quantitatively expressed. Each set of experiments usually compares the various specimens by subjecting them all to the same treatment at once, and including, as a standard, specimens of stone or brick of known value.

The usual method of conducting this test is to place whole bricks in a foundry tumbler with a given weight of cast-iron, and determine the loss of weight of the specimens after a certain number of revolutions. It is customary to repeat this operation for two or three independent periods, usually about one half hour each. The loss during the first period is largely influenced by the chipping of corners of the bricks, and the test of wear would be based more upon the later periods. The results of such tests made at different times and places are not comparable with each other, but they are useful as showing the comparative merit of the samples at hand in each case.

Where specifications include a test of this character, it is better to require that the brick shall bear a certain relation to a standard material to be included in the test rather than to name a minimum percentage of loss, as there is necessarily considerable uncertainty in the test, unless it can be conducted according to a method the results of which upon standard material are definitely known, and the details of which may at any time be reproduced both as to the apparatus to be used and as to the nature, sizes and forms of the abrading material. It would be advantageous if certain standards could be recognized and used in such work which would enable comparisons to be made of different material upon the basis of percentages obtained by various men in different localities.

It is also desirable to use pieces less in size than a whole brick, in order that the abrasion during the test may not be altogether upon the outside of the brick.

It is quite true that the action to which the material is subjected in a test of this character is quite different from the wear to which the material is subjected when firmly held in the pavement, but the qualities necessary to resist wear in the two cases are quite similar. We may form an idea of whether a material is suitable for the proposed use from such experiments, although no definite idea of the amount of wear that it will endure can be obtained from them.

Absorption tests are made by weighing the specimen dry, then saturating it and weighing again, and stating the absorption as a percentage of the dry weight. In making the test, the brick is first thoroughly dried by placing it in a drying oven at a temperature of 212° Fahr. until it ceases to lose in weight. It is then placed in water and permitted to remain until saturated and weighed again. In some cases the brick is left in the water twenty-four hours, in others until it ceases to materially gain in weight. The latter is preferable, as the absorption of various bricks may differ not only in amount, but also in rate. Whole bricks should not be used for absorption tests, as the outside is likely to be less absorptive than the interior. A good paving-brick will not usually absorb more than 5 per cent of water, and many of the best varieties will take less than 1 per cent.

Tests for the presence of free lime may be made by placing a specimen in water and leaving for a few days. If uncombined lime be present in considerable quantity it will cause the brick to crack or blow on the surface. Tests for this purpose may also be made by pulverizing a small portion of the brick, washing it with water, and determining the percentage of soluble matter contained by the brick.

Such tests may aid in forming a judgment as to the value of a material for paving purposes, the only conclusive test, however, is the record of use of the same material in similar work. If the normal value of a certain make of brick be known, tests may indicate whether a special lot be of standard quality. They cannot be conclusive as to the value of an untried material.

ART. 53. FOUNDATIONS FOR BRICK PAVEMENTS.

A brick pavement should have a firm foundation. As the surface is made up of small independent blocks, each brick must be adequately supported from below, or the loads coming upon it may force it downward and cause unevenness. The wear of the pavement depends very largely upon the maintenance of a smooth even surface, as any unevenness will cause the bricks to chip on the edges, and also produce impact from the loads passing over the pavement.

The best foundation for a brick pavement is doubtless one of concrete, laid after the manner given in Art. 47. For light or moderately heavy traffic, such as that of the ordinary small city, the concrete is usually placed 6 inches thick. If the traffic be very heavy 9 inches may be necessary, and where from any cause the road-bed is not firm it may be advisable to still farther increase the depth.

Under comparatively light traffic a foundation of gravel or broken stone as mentioned in Art. 46 may be used. This foundation should, however, usually be employed only where traffic is light and the road-bed good.

The double-layer pavement (see Fig. 22) consists of a foundation made by placing a layer of sand or gravel 3 to 5 inches thick upon the road-bed, rolling it thoroughly and laying a course of bricks upon it. The bricks are laid flat with their greatest dimension lengthwise of the street, as explained in Art. 48.

This foundation has been more extensively used under brick pavements than any other, and has often given satisfactory results. It is now largely giving place to concrete in the better class of work, and in many cases under light traffic its economy is questionable, as the layer of gravel would often answer equally well without the lower layer of bricks. The best base to use for a particular work must usually be largely determined by the availability of various materials.

ART. 54. CONSTRUCTION OF BRICK PAVEMENT.

In laying a brick pavement, after the completion of the foundation as described in Chap. VI and Art. 53, a cushion coat of sand is spread over the surface of the foundation, $\frac{3}{4}$ inch to $1\frac{1}{2}$ inches thick, to receive the surface layer of bricks. The cushion coat should be composed of clear sharp sand, and quite dry when the surface brick are placed upon it. It is also desirable that the sand layer be rolled with a light roller and brought carefully to surface, in order that it may afford firm and even bearing for the brick.

The brick surface should be composed of carefully selected material, as durable, impervious, and uniform as possible. The bricks should be laid on edge, in close contact with each other. They are usually arranged in courses at right angles to the line of the street, the greatest length of the brick being across the street, and the bricks in adjoining courses breaking joints with each other. The laying is begun at the curb, alternate courses beginning with whole and half bricks and working to the centre, the work at the curb being carried but very little ahead of that at the middle of the street, in order that partial courses may not be disturbed before being completed across the street. This system of construction is shown in Fig. 21, which represents a pavement as constructed for heavy traffic on heavy concrete foundation.

In many cases the gutter-bricks are turned with the

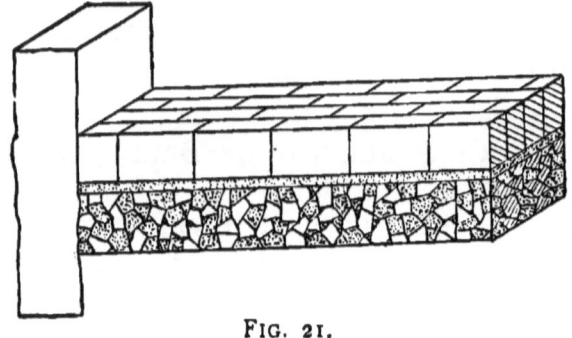

FIG. 21.

greatest dimension lengthwise of the street, with the object of facilitating the flow of surface-water in the gutter. The advantage of this is doubtful, as it has the effect of breaking the bond of the pavement between the gutter-bricks and roadway. This is shown in Fig. 22, which shows the construction of a double-

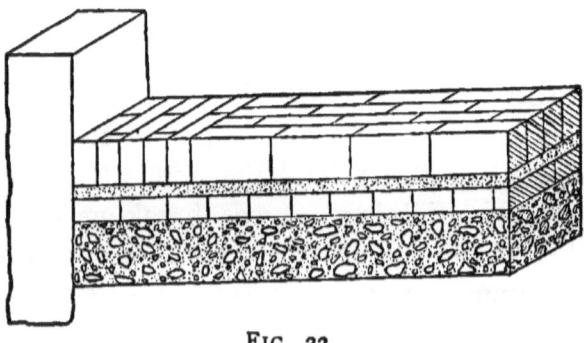

FIG. 22.

layer pavement with brick and gravel base, as has been commonly used under light or moderate traffic.

Some engineers advocate laying the brick at an angle

of 45 degrees with the street line, except on streets where there are street-railway tracks. Mr. Niles Merriwether, City Engineer of Memphis, Tenn., in his report for 1893, expresses the opinion, based upon experience in that city, that this arrangement of bricks conduces to good wear in the pavement.

After the surface layer of brick is in position it should be rammed or rolled to a smooth and uniform surface. Usually a heavy wooden rammer is employed for this purpose, and the ramming should be so thoroughly done as to discover any weak places that may exist in the pavement by forcing the bricks down out of surface at such points. When such places are discovered the bricks should be removed and the sand filled in below to properly support them.

When the bricks are well rammed and brought to the proper surface the joints should be filled with material of an impervious nature that will cement the bricks together and form them into a solid and impervious roadway. Various mixtures of coal-tar and asphalt are commonly used for this purpose. A grouting of hydraulic cement mortar is also sometimes employed for this purpose.

After the joints are well filled a coating of sand $\frac{1}{4}$ or $\frac{1}{2}$ inch thick is given to the pavement and it is opened for traffic. In some cases the entire surface of the pavement is coated with the tar, and the layer of sand is applied hot, with the view of insuring the binding of the surface-bricks and rendering the pavement less pervious.

In many cases also the tar filling is omitted altogether and the surface after ramming is covered with the sand layer and opened to travel.

There is considerable variation in the methods of construction employed in different localities, as to the kind of foundation used, the arrangement of the surface-bricks, and the filling of the joints. A single-course pavement on a light gravel foundation with joints filled only with sand or small gravel has frequently been used for light traffic under favorable conditions with satisfactory results. Solid and impervious construction, however, will always give the best results in wear, and usually will be most economical in the end.

ART. 55. MAINTENANCE OF BRICK PAVEMENTS.

The maintenance necessary for a brick pavement consists in keeping it clean and carefully watching it, especially during the first year or two years, to see that no breaks occur due to the use of defective bricks in the surface or to insufficient support from the foundation at any point. When any unevenness from either of these causes appears, it should be at once rectified before the pavement becomes irregularly worn in consequence.

While, as already stated, the utmost care should always be taken to use only material of a uniform quality in the surface of the pavement, still under the closest inspection some inferior material may be used, which will only be shown when wear comes on the pavement, and unless then removed at once it will cause the evenness of the surface to be impaired about it. Irregular support from the foundation will be less likely to occur in good construction, but its effect will be similar to defective material, the sinking of individual bricks producing uneven wear. Weak spots in the foundation may sometimes be caused, where concrete foundation is not employed, by surface-water which is permitted to pass through the joints, saturating the sand or gravel beneath and causing it to move under concentrated loads. For this reason the joints should be observed during the early wear of the pavement in order to remedy any case where they may not have been properly filled.

Where a brick pavement has been constructed of good material and kept in good surface during the early period of use, it may then reasonably be expected to wear out without any considerable expense for small repairs. The length of time the pavement may be expected to wear depends upon the quality of the material and the methods of construction. For the moderate traffic of many of the smaller cities, and lesser streets in the large cities, brick has shown an endurance which indicates it to be a satisfactory and economical material, and it is not improbable that by careful attention to proper construction it may be used for even heavier traffic.

CHAPTER VIII.

ASPHALT PAVEMENTS.

ART. 56. ASPHALTUM.

ASPHALTUM is a mineral pitch which occurs in a number of localities widely distributed over the surface of the earth. It is supposed by most authorities to be the result of the decomposition of vegetable matter, although by some it is considered to be of volcanic origin. It consists, in its natural state, of bitumen with a small percentage of foreign organic matter and mixed with more or less mineral earth, and varies according to the nature of the bitumen from the soft viscid condition of mineral tar to a hard brittle substance of glassy appearance and conchoidal fracture. It also occurs in a solid state as a rock impregnated with bitumen, which will be separately considered in another article under the head of *rock asphalt*.

The asphaltum used for street pavements in this country is obtained for the most part from the island of Trinidad, W. I., and from the state of Bermudez, Venezuela. This asphaltum is known as *lake asphalt* or as *land asphalt* according to the source from which it is derived. *Lake asphalt* is found in large deposits known as the pitch lakes. The lakes cover a considerable area,—in Trinidad about 100 acres; in Bermudez several hundred,—and the pitch seems to be, or have been, forced upward from below through fissures in the rock or craters. The pitch upon contact with the air gives off gas and gradually hardens. In the lakes proper the asphaltum is more or less in motion, and excavations in the surface are soon filled by the flow of material from the sides and bottom. The pitch near the centre of the lake at Trinidad is more soft than near the sides, and it has been supposed that the supply from subterranean sources still continues to some extent. It has also been found that the surface of the lake is higher in the centre than at the sides, and that the general elevation of the surface has been lowered somewhat by the large quantities of material which have been removed from it.

The so-called *land asphalt* from Trinidad is found in vicinity of the lake, and is a harder material than the lake asphalt, probably from longer exposure to the air. It may have been derived either from overflow of the lake or from independent subterranean sources the action in which has now ceased. (For a complete description of the Trinidad pitch deposits see the "Report of the Inspector of Asphalts and Cements of the District of Columbia" for 1891–92.)

Asphaltum similar in character to that already mentioned occurs at many other places. Mines are worked in Cuba and Peru, and large deposits are found in Mexico. It occurs at a number of places in Europe, while the bituminous mortar of the ancient Chaldean constructions was of this character, and beds of asphaltum are still found in that country from which it may have been obtained. In the United States deposits of asphaltum are found in California, Utah,

West Virginia, and other places. Those of California have to some extent been applied to paving purposes, but have not as yet been largely employed.

The crude asphaltum usually contains considerable water as well as earthy and vegetable impurities. It is heated in a boiler to 300° or 400° Fahr., the water being driven off, and the impurities settling to the bottom or forming a scum on top. The liquid asphaltum is then drawn off and is known as *refined asphalt*. This refined asphalt may contain more or less mineral or earthy matter distributed through it in a finely divided state. Refined Trinidad asphalt may contain 52 to 56 per cent of pure bitumen, while Bermudez asphalt is said to contain 97 per cent.

The refined asphalt is brittle at ordinary temperatures, and possesses little cementitious value. To reduce it to a condition from which it may be easily compacted in the pavement, it is heated to a temperature of about 300° Fahr. and mixed with the oil residuum obtained from the distillation of petroleum. This mixture is very thoroughly worked, so as to form a material of uniform consistency. The product is then known as *asphaltic paving cement*.

Great care is necessary in mixing the paving cement to properly proportion the ingredients, as the value of the cement depends upon their nice adjustment. Both materials are quite variable in their properties, and continual tests are necessary of the materials employed as well as of the resulting product, in order to obtain a cement of uniformly good quality. The quantity of oil necessary varies with the nature of the asphaltum, being less as the asphaltum is of a more plastic and less refractory nature.

The surface material for a pavement of asphaltnm is formed by mixing asphaltic paving cement, prepared as already described, with sand, or with sand and powdered limestone. The proportions of the various ingredients depend upon their character and upon the requirements of the pavement. In order to obtain good results it is necessary that the exact nature of each of the ingredients be determined, and the proper amounts used. With Trinidad asphaltum the surface material ordinarily contains about 10 per cent of pure bitumen.

The sand used for this purpose is usually very fine, 60 per cent being sometimes required to pass a sieve of 60 meshes per linear inch, and all of it through one of 30 meshes. It is important also that the sand be clean and free from loam or clay. The carbonate of lime is used in the form of fine powder.

In forming the surface material the sand and paving cement are separately heated to a temperature of about 300° Fahr. and mixed, while hot, in an apparatus which thoroughly incorporates them into the mixture. When powdered limestone is used it is added cold to the hot sand before mixing with the hot paving cement.

As may be readily seen, the selection of asphaltum for paving purposes, as well as the process of forming the surface material, is a matter requiring very great care and an intimate knowledge of the characteristics of the materials to be employed. In order to secure good results it is quite essential that careful examination be made of every mixing of the material for the surface, both by testing the materials before they are used and the product after it is formed. Analysis of every lot should be made as well as consistency tests. Experience is the only guide, and there is no method of judging of value in such work other than by the results of former work of the same kind.

Variations in the quality of asphaltum are attributed to the character as well as the quantity of bitumen that it contains. This bitumen is made up of two parts, the first of which, called petrolene, is the oily and cementitious material; the other, called asphaltene, is the hard material lacking, in the cementing properties. The varying proportions of these ingredients in the bitumen determine the temperature at which it will melt, and the facility with which it may be used for paving purposes. A certain proportion of asphaltene is probably necessary to give the material sufficient stiffness at ordinary temperatures; but too large a quantity makes it lacking in plasticity and cementing properties, and renders it necessary to use a large proportion of residuum oil in preparing the asphaltic cement.

The residuum oil also varies considerably in character, and the quantity to be used in each case depends somewhat upon the character of the oil. The object aimed at is to get a cement of given consistency, which is measured by the penetration of a needle, at a standard temperature and under a standard weight. In forming the paving cement it is necessary that the materials be constantly agitated in order that the oil and asphaltum may be intimately mixed, and that the mass of cement may be uniform throughout.

For making a paving cement, in some cases where American asphalt has been used, a maltha, or liquid bitumen, has been substituted for the residuum oil to reduce the asphaltum to plasticity. This maltha is a nearly pure bitumen, similar in character to the asphaltum, but differing in that the bitumen contains a high percentage of the hydrocarbon known as petrolene and but very little asphaltene, the proportion between the two being such as make the bitumen liquid at ordinary temperatures.

The composition of surface material for asphalt pavements must be varied to suit the conditions under which each is built. The variations of temperature to which the pavement may be subjected are of special importance, and the nature of the traffic may also have an influence. The surface must not soften under the heat of summer, and yet must be sufficiently plastic not to become brittle and chip off in cold weather. For light traffic the material may be more soft in warm weather than under heavy traffic, as it is not so liable to cutting under the loads.

The quantity of paving cement used in the surface material may depend somewhat upon the consistency adopted for the cement. If it be stiff a larger percentage may be used, with the same effect as to the softening of the surface in summer than if it be soft. The common proportions of ingredients employed in forming the surface material are approximately as follows:

	For Trinidad Asphalt.	For Bermudez Asphalt.
Asphaltic cement	12 to 15 per cent.	9 to 10 per cent
Sand	83 to 70 "	71 to 60 "
Carbonate of lime	5 to 15 "	20 to 30 "

In some cases stone-dust, formed by crushing a hard stone, such as granite, to about the dimensions of a fine sand, is substituted for a portion of the sand and carbonate of lime.

The proper treatment in any instance can only be determined by a careful study of the materials to be used, the climatic conditions, and the service required.

Art. 57. Rock Asphalt.

Rock asphalt, as commonly used in paving, consists of limestone naturally impregnated with bitumen in such proportion as to form a material which may be softened by the action of heat and again consolidate when cooled if brought under pressure.

This rock is mined at several places in Europe, notably at Seyssel, France; Travers, Switzerland; Ragusa, Sicily, and Vorwohle, Germany. It is usually composed of nearly pure carbonate of lime, impregnated with from 7 to 20 per cent of bitumen. It occurs in veins, after the manner of coal, is hard at the ordinary temperatures of the mines, and is quarried by the use of explosives.

The preparation of the surface material of rock asphalt consists only in crushing and grinding the rock to powder and heating the powder to drive off the water and soften it, so that it may be compacted in the roadway. The powder is heated to a temperature of from 200° to 300° Fahr., and is applied while hot in laying the surface.

Natural rock asphalt, suitable for paving purposes, usually contains from 9 to 12 per cent of bitumen. If it contain much more than this it is apt to become soft in warm weather. If it contain less it will not consolidate properly or bind well in the pavement.

The rock should be of fine even grain, and have the bitumen uniformly distributed through it. In forming the surface material for rock asphalt pavements, the rock from different mines is commonly mixed in such proportions as to give about 10 or 12 per cent of bitumen to the mixture, thus making a harder surface than would be obtained by the use of the rich rock alone, as well as less likelihood of softening. No other material is mixed with the rock in forming the surface.

In determining a mixture of asphalt rock, as in the use of the lake asphalt, the local conditions of climate and traffic must be considered, and the quantity of bitumen be so proportioned as to remain solid in summer and not become brittle and lose cohesion in winter. Experience with the material and exercise of great care in the determination of proper proportions is therefore essential to success in the construction of any asphalt pavement.

In the use of bituminous limestone for sidewalks and many other purposes where a plastic material is required, the rock asphalt powder is mixed with an additional quantity of bitumen, or asphaltum, sufficient commonly to give a product containing 15 to 18 per cent of bitumen. This product is known as *asphalt mastic* in Europe. For use in sidewalks the mastic is melted and mixed with sand or gravel to form a wearing surface.

A sandstone impregnated with bitumen occurs at a number of localities in the United States. This stone has been applied to some extent in paving, in somewhat the same manner as the European material. It is still in the experimental stage, and has not come into general use.

ART. 58. ASPHALT BLOCKS.

Asphalt paving blocks are commonly formed of a mixture of asphaltic cement with crushed limestone. This limestone is crushed to sizes of $\frac{1}{4}$ inch in diameter or less, and mixed with the asphaltic paving cement formed as described in Art. 56, in such proportions as that the product contains about 10 per cent of bitumen.

The materials are heated to a temperature of about 300° Fahr., and mixed while hot in an apparatus arranged to secure the even distribution of the ingredients through the mass. The thorough incorporation of the various materials in the mixture is of first importance in producing homogeneous and uniform blocks, while the quality of the materials used needs as careful inspection as in the case of the surface material for sheet pavements.

When the mixing is complete, the material is placed in moulds and subjected to heavy pressure, after which the blocks are cooled suddenly by plunging into cold water.

These blocks have usually been made larger than paving-bricks, the common size being 12 inches long, 4 inches wide, and 5 inches deep. They are laid in the same manner as brick, as closely in contact as possible, and driven together. Under the action of the sun and the traffic, the asphalt blocks soon become cemented together, through the medium of the asphaltic cement, and form, like the sheet asphalt pavements, a practically imperivous surface.

They are commonly used on streets of light traffic only, as the blocks as heretofore constituted wear rapidly under heavy traffic. They are usually laid upon a foundation of sand or of sand and gravel, and on account of the impervious nature of the surface may often give satisfactory results on such a foundation, where a more pervious block pavement or a sheet pavement would require more efficient support.

A large amount of pavement, of blocks made practically as described above, has been laid in this country, particularly at Baltimore and Washington, and have shown good durability in wear under moderate and light travel.

These blocks have the advantage over sheet asphalt for the smaller cities, that the blocks may be formed at a central-point and shipped ready for use to the site of the proposed pavement, and that no special plant need be erected in each town where they are to be constructed.

In forming the asphalt block pavement the road-bed is brought to subgrade in the ordinary manner and rolled, leaving room for the pavement of uniform thickness to be placed upon it. A layer of gravel 4 or 5 inches deep is then placed and rolled, with a cushion coat of sand 1 to 2 inches, and then the paving blocks. The blocks are pressed together in the courses by the use of a lever, and the courses driven against each other with a maul to reduce the joints as much as possible. A coating of sand is given to the surface of the pavement, and it is rammed to a firm and uniform surface, as in the case of brick.

ART. 59. FOUNDATIONS FOR SHEET PAVEMENTS.

As a sheet asphalt surface has no power to sustain loads, acting only as a wearing surface, which must be held in place from below, it is essential that it be placed upon a very firm, unyielding foundation. It is consequently nearly always placed upon a concrete base, which is commonly formed of hydraulic cement mortar and broken stone, prepared as described in Art. 47. In the use of this base, it is necessary that the mortar be fully set, and the concrete thoroughly dry before the asphalt is laid upon it, as the placing of the hot surface material upon a damp foundation will cause the blistering and possible disintegration of the surface by the steam generated from the base by the heat of the material.

For moderate or heavy traffic in cities, the concrete base is commonly made 6 inches thick. For lighter traffic a less depth, 4 inches or 5 inches, is sometimes employed. The depth necessary will depend upon the nature of the road-bed as well as the weight of the traffic. It should be greater as the subsoil is less firm and well drained.

Frequently the concrete for the foundation is formed of asphaltic or coal tar paving cement instead of hydraulic cement mortar. It is then known as *bituminous concrete*, and the foundation is called a *bituminous base* to distinguish it from the ordinary *hydraulic base*. The advantage claimed for the bituminous base is that the foundation and surface material become joined into a single mass, with the effect of anchoring the surface and preventing the formation of weather-cracks and wave-surfaces, which sometimes occur when the hydraulic base is used, under a light surface layer, in consequence of the lack of bond between the hydraulic concrete and the asphalt surface. Another reason for the use of this base is that it is cheaper than the hydraulic base.

An intermediate layer known as the binder course is now commonly added to the foundation, or rather placed between the foundation and surface layers. This layer is usually formed of coal-tar or asphaltic paving cement, mixed with small broken stone, not more than 1 inch in diameter, about one gallon of the cement being required to 1 cubic foot of stone.

The materials are mixed hot and laid and rolled in the same manner as the surface layer. This binder becomes consolidated with, and gives added depth and strength to, the surface; thus having a tendency to prevent the cracks and wave-surfaces which may otherwise appear in the surface when used upon an hydraulic base.

The hydraulic base is usually preferred to the bituminous base on account of its forming an unyielding structure, not likely to be forced out of place by the weight of the traffic at any point where the support of the road-bed may be weakened, while the bituminous concrete has not the strength to resist deformation under heavy loads unless uniformly supported. The joining of the base and surface into one mass, as is effected by the bituminous base is also a disadvantage when the pavement is to be resurfaced, as with the hydraulic concrete base the surface may be easily stripped off, and a new surface placed without injury to the foundation.

ART. 60. CONSTRUCTION OF SHEET PAVEMENTS.

The construction of asphalt pavements in this country is, in the main, in the hands of two or three large corporations, and methods of construction vary but little, the differences in the various pavements being principally due to differences in the composition of the materials used.

In constructing the pavement after the completion of the foundation as indicated in the preceding article, the surface material is brought to the place where it is to be used in a large kettle and applied in a hot, semiplastic condition and thoroughly consolidated by rolling. The tools with which the material is handled are kept hot, hot rakes being ordinarily employed for spreading it, and hot rollers for the first compacting.

Sometimes, as already indicated, a binder course is inserted between the base and surface layers. This binder course is usually $1\frac{1}{2}$ inches thick, composed of coal-tar cement and small broken stone, and applied in the same manner as the surface layer. This construction has been considerably used in Washington, and a pavement so constructed is known as the *combination pavement*.

In some cases where the binder course is omitted, the surface is applied in two layers, of which the lower, known as the cushion coat, is made from $\frac{1}{2}$ inch to 1 inch thick when compacted, and contains a higher percentage of paving cement than the surface layer. This layer having more of the cementitious material adheres more strongly to the hydraulic base, and forms a tie between foundation and surface.

When a bituminous base is employed the cushion

coat is generally considered unnecessary, as the surface layer will join directly with the base. The binder course is, however, frequently used in this case to give added strength and weight to the pavement.

The surface coat, prepared as described in Arts. 56 and 57, is usually applied so as to be about $1\frac{1}{2}$ to $2\frac{1}{2}$ inches thick when compacted in the pavement. When a binder course is employed about $1\frac{1}{2}$ inches, otherwise 2 inches to $2\frac{1}{2}$ inches.

In the construction of rock asphalt pavements, which have been almost exclusively used in Europe, it is common to ram the surface layer with hot rammers, and smooth it off with smoothing-irons, while in this country small hot rollers have usually been employed. The practice of the rock asphalt companies has also differed from that of those using lake asphalt, in that the latter continue rolling the surface with heavy rollers until it is hardened and shows no mark, while the European practice is to roll more lightly and leave the final compression to be given by the traffic.

In Europe foundations of hydraulic cement concrete are exclusively used, with the surface layer usually directly in contact with it. This system is also most commonly used in this contry, the standard pavement being formed of an hydraulic base 6 inches thick with a single surface layer $2\frac{1}{2}$ inches thick when compacted, or where lighter construction is admissible, an hydraulic base 4 inches thick is used with a surface layer 2 inches in thickness.

In all asphalt pavements it is customary during the rolling to give the surface a coating of hydraulic cement, which is usually swept lightly over the surface.

Art. 61. Vulcanite or Distillate Pavement.

Numerous attempts have been made to construct a pavement by the use of coal-tar as a cementing material in place of or in conjunction with asphaltum. But few of these have met with any degree of success. Coal-tar by itself, as most commonly employed, is soon disintegrated by the action of the weather; it is also strongly affected by temperature, becoming soft in hot weather and brittle in cold weather.

Pavements in which the wearing surface is composed of a mixture of coal-tar and asphaltum have in some cases given good results in practice. These pavements are known as vulcanite or coal-tar distillate pavements. They are much cheaper than asphalt. They are said to be somewhat less slippery, and to resist better where exposed to dampness. The vulcanite surface, however, is not so durable under wear, and it requires very great care in construction to produce a surface of uniformly good quality, because of the possible variations in the nature of the coal-tar used.

In the preparation of the surface material for distillate pavement as it has been used in Washington, a paving cement is used containing 70 to 75 per cent of coal-tar paving cement and 25 to 30 parts refined asphaltum. The construction of the pavement is in other respects similar to that of the combination asphalt pavement on a bituminous base. As commonly constructed the pavement includes a 4-inch bituminous base, a $1\frac{1}{2}$-inch binder-course, and a surface of tar distillate $1\frac{1}{2}$ inches thick. The material for the wearing surface for this pavement, according to the specifications of 1892, was formed as follows:

ASPHALT PAVEMENTS. 149

Clean sharp sand..........63 to 58 per cent.
Broken stone or rock dust..23 to 28 " "
Paving cement............13 to 15 " "
Hydraulic cement.. ...... 0.9 " "
Slaked lime............... 0.15 " "
Flour of sulphur.... ...... 0.1 " "

The materials are heated to 250° Fahr. and mixed hot, then laid after the manner of asphalt.

ART. 62. MAINTENANCE OF ASPHALT PAVEMENTS.

To give good service asphalt pavements must be kept clean. On account of the smooth surface and absence of joints, cleaning may be readily accomplished; and the presence of dirt, especially in wet weather when it is likely to cause the surface to remain damp, is liable to cause the asphalt to rot. More than any other pavement, therefore, the durability and wear of an asphalt surface depends upon its cleanliness. The presence of dirt upon asphalt in damp weather is also important in its effect upon the slipperiness of the pavement.

Small repairs of any breaks that may occur in an asphalt surface may be easily made, and such repairs should be constantly attended to in order to keep the surface in good condition. Small breaks will rapidly extend if they are not repaired at once. In making repairs to the surface of the pavement it is necessary to cut away the surface for a short distance about the imperfect spot, stripping the surface from the foundation and cutting the layer down square at the edges, after which a new piece of surface may be introduced to fill the hole in the same manner that the original surface was constructed. Such a patch may ordinarily be put on so as to make joints that will join perfectly with the old pavement and not show where it has been placed. When a surface has become so worn that patches would be numerous, the old surface may be stripped off and a new one placed upon the original foundation. When repairs are to be made upon a pavement having a bituminous base it is more difficult to cut out the holes in satisfactory shape, as there is no well-defined joint between the base and the surface layers.

The repairs that may be required upon an asphalt pavement depend, of course, upon the solidity of construction and the nature of the surface material. There is so great variation in the materials employed for the wearing surface that, as would naturally be expected, very considerable difference in wear is shown by different pavements. Asphalt pavements for the most part, as has been stated, are built by corporations employing corps of expert workmen, and the plan usually adopted is to require the contractor to keep the pavement in repair for a certain length of time without charge and afterwards to maintain it for a certain longer period at a fixed annual price. This makes it an object for the contractor to do good work, and is the most effective way of securing it where so many elements of uncertainty enter.

CHAPTER IX.

WOOD PAVEMENTS.

ART. 63. WOOD BLOCKS.

WOOD for pavements should be close-grained and not too hard. It should be as homogeneous as possible in order that the wear may be uniform, and soft enough that it may not wear smooth and slippery. To give good service in wear the wood should be penetrated by water as little as possible and show good resistance to decay under the action of the weather.

Wood for this use should be sound and well seasoned. The blocks should always be subjected to careful inspection. All sapwood needs to be removed in order to lessen the liability to early decay, and blocks containing shakes and knots should be rejected.

In the United States cedar has been most largely used for this purpose and has proved to be a quite satisfactory wood for such use. Yellow pine and tamarack have also been employed for pavements to some extent at the North, and cypress, juniper, cottonwood, and mesquite at the South. These varieties have all been used with some success, and can be made to furnish a fairly good paving material.

Oak and other hard woods are less likely to wear evenly in the pavement, become smooth and slippery, and bear less well the exposure to the weather and influences tending to cause decay in the wood. In Washington, D. C., a pavement of hemlock blocks was at one time constructed on quite an extensive scale, but proved unsatisfactory and was soon destroyed. It does not appear, however, to have been a well-constructed pavement or of properly selected material.

In Australia hard-wood blocks have been quite extensively used and are reported as giving good service, although they are admitted to be somewhat slippery in wet weather. Australian *Karri* and *Jarrah* woods are employed, and it is claimed for them that they show unusually great resistance to wear and are not soon affected by decay.

In London, where wood pavements have been very extensively employed, Swedish yellow deal is commonly placed at the head of the list of woods in value, yellow pine and Baltic fir being also largely used and considered good in use. The Australian woods above mentioned have also been used to some extent in London, and are said to have given very satisfactory service, showing greater resistance to wear than deal or pine, although somewhat expensive.

Wood pavements are commonly constructed of blocks set with the fibres vertical, so that wear comes upon the ends of the fibres and has no tendency to split pieces off from the blocks. Cedar blocks are used in the form of whole sections of the tree, on account of the liability of the wood to split off between the layers when cut to a rectangular shape. They are usually of an approximately cylindrical form, varying from 4 to 9 inches in diameter and 5 to 8 inches in depth. Commonly the whole section is used with the bark removed, giving blocks somewhat irregular in shape. In some cases the blocks are, however, cut to a true cylindrical form, the sapwood as well as the bark being cut away, by passing them through a set of knives which are gauged to turn out cylinders of given size. Several sets of knives are employed, giving blocks of varying sizes, each block being cut sufficiently to eliminate all sapwood with as little waste of good material as possible. The use of sapless blocks increases the life of the pavement by augmenting the resistance of the material both to wear of the traffic and to the disintegrating influences of the atmosphere and moisture.

Wooden paving-blocks other than cedar are usually of rectangular form, 8 to 12 inches long, 3 to 5 inches wide, and 5 to 8 inches deep. As the blocks are usually laid in courses, the width as well as the depth must be constant for the same work. It is usual to cut the blocks from plank of uniform thickness,—ordinarily 3 inches,—as the narrow blocks give a better foothold for horses in damp weather, and also are more easily settled to a firm and even bearing in the pavement.

Hexagonal blocks are also sometimes used. In a mesquite pavement constructed at San Antonio, Texas, the blocks are hexagonal in form, the tops being slightly smaller than the bottom, the diameter varying from 4 to 8 inches, with a depth of 5 inches.

The tendency of recent practice in constructing wood pavements has been in the direction of making the depth less than was formerly used. Five inches is now a very common depth.

Deep blocks are usually a waste of material, as in most cases not more than 2 or $2\frac{1}{2}$ inches at most can be worn from the pavement before it is replaced, even if the traffic be sufficient to wear it out before it rots.

In the use of the Australian hard woods in London a less depth of block has been tried and found satisfactory, and a depth of about $3\frac{1}{2}$ inches is recommended by some of the engineers.

Art. 64. Foundations for Wood Pavements.

The foundation which has been most commonly used in the United States for wood pavement is that composed of a layer of boards upon sand, as described in Art. 49. This foundation has been used in a number of places with fairly good results, and, under light traffic, where the first cost of the pavement must be low, its use may sometimes be advantageous. This form of construction has the disadvantage of being less firm than the others usually employed, as well as that of being perishable in its nature.

The life of the pavement is therefore likely to be less than upon a more firm and durable base, as the destruction of the surface of the pavement may be caused by the yielding of the foundation. There have been instances recorded where a base of this kind has worn out two or more surface layers, and there are others in which the failure of the foundation planks is the beginning of the destruction of the surface. Much depends upon the locality and the quality of the work.

On the better class of wood-block pavements, a foundation of concrete is usually employed. This gives firm support to the blocks, and admits of even wear upon the surface of the pavement. A durable foundation also has the advantage that when the surface layer is worn out, the pavement may be resurfaced without renewing the foundation.

In using a concrete foundation, a cushion coat of sand is commonly employed on top of the concrete in which to bed the blocks in order that they may be brought to an even surface. Sometimes a thin layer of cement mortar is used in place of the sand upon the concrete; and in London some pavements have been constructed with a thin layer, about $\frac{1}{2}$ inch, of asphalt mastic over the concrete, the blocks resting upon the mastic.

Broken-stone and gravel bases are also frequently employed under wood blocks, a layer of boards being usually placed over the broken stone to form an even bearing for the bases of the blocks, although sometimes the blocks are bedded in a cushion coat of sand, as with the concrete foundation. In Duluth, a Telford foundation has been employed under wood, laid in the same manner as for a broken-stone road. A layer of gravel was placed upon the Telford pavement and rolled to a smooth surface, after which the wood blocks were laid directly upon the gravel. The reason given for the application of this foundation instead of concrete in this instance is that the road-bed is of soft material, which in many places could not be compacted by the use of a heavy roller before the placing of the concrete, while the stones of the Telford foundation are forced to a firm bearing and give a uniform support. This difficulty of a soft road-bed has been sometimes met in other places by rolling a thin layer of gravel or broken stone into the surface of the road-bed, the forcing of the stones into the soil causing it to become compact and firm, after which the concrete may be placed in the usual manner.

It is perhaps even more important that the foundation for a wooden pavement be firm and unyielding than with other kinds of block pavement. Any small motion due to the flexibility of the base is likely to split the blocks, and if through the yielding nature of the foundation some of the blocks are forced out of surface so that the surface becomes slightly uneven, the wear will be very greatly increased over what would be the case if the foundation were firm and immovable.

ART. 65. CONSTRUCTION OF WOOD PAVEMENTS.

In the construction of pavements of cylindrical blocks, or whole-tree sections, as in the common cedar pavement of the United States, blocks of varying sizes are usually employed, being set in contact with each other, the smaller blocks between the larger ones, in such a way as to leave the spaces between blocks as small as possible. With blocks varying from 4 to 8 inches in diameter, this arrangement gives good foothold for horses, and at the same time reduces to a minimum the wear due to the blows caused by wheels sinking into the joints where large spaces exist between the blocks.

The common arrangement of cedar blocks is shown in Fig. 23, which represents a section of pavement as ordinarily constructed on a sand and board foundation.

In the use of rectangular blocks, the blocks are set with their longest dimension transverse to the length of the street. They are usually arranged in courses across the street, being placed close together in the courses, and arranged to break joints in adjoining courses. Between courses a joint is usually made $\frac{1}{4}$ to $\frac{1}{2}$ inch in width, for the purpose of affording a foothold to horses. In the older pavements of this character a much wider joint was employed, some as much as an inch in width, with the idea that they were necessary to secure proper foothold. Experience has shown, however, that the wide joints are not necessary; and it

FIG. 23.

is now commonly agreed that where, as is now common, the blocks are limited to a width of 3 inches, a joint ¼ inch in width is sufficient for the purpose. As in the case of round blocks, durability is greater where the joints are as small as possible, and the liability of the fibres of the blocks to spread is eliminated.

The tendency in practice is toward a continual diminution of the width of the joint, and some pavements have been constructed in London in which the courses of blocks are placed in contact with each other. These are reported to have given good results in service, and to be advantageous in increasing the durability of the pavement. In some cases, where the pavements are laid with close joints, an expansion joint ¼ inch in width is provided for every 30 inches of length, in order to provide for the swelling of the blocks. In other cases, however, this is said to have been omitted without injury to the pavement.

The method of setting rectangular blocks is shown in Fig. 24, which represents a wood-block pavement on a concrete foundation, as commonly constructed.

In laying a pavement of this kind, a course of blocks is first set across the street, and then a strip of wood of the thickness of the joint is set against the row of blocks and left until the next course is placed, or

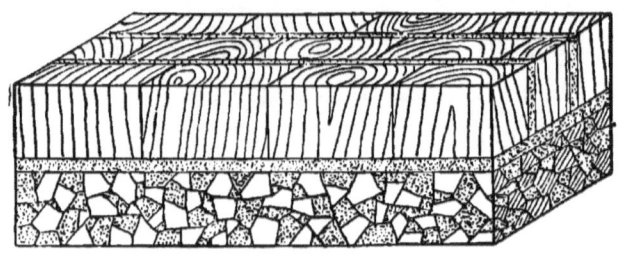

FIG. 24.

sometimes spuds with heads of the thickness of the joints are driven to the head in the side of each block, and the next row of blocks are set against the spuds.

Where blocks are set in courses across a street, it is necessary that allowance be made at the curb or gutter for the expansion of the blocks. This is usually accomplished by leaving an open or sanded joint until the blocks have done swelling.

The gutter blocks are very commonly turned with their lengths along the street, and sometimes the course next the curb is left out temporarily and filled with sand to provide for swelling.

Various methods are employed for filling the joints between the blocks. It has been the common practice in the construction of the cheaper wood pavements in the United States to fill the joints with sand and gravel, sometimes with a coating of tar, or in some cases the

WOOD PAVEMENTS. 159

joint is partially filled with tar and then completely filled with sand or small gravel. The objection to this method is that it does not make an impervious joint, which in a wood pavement, and especially with the cheaper foundations, is a matter of the utmost importance.

The best practice seems to be to fill the lower part of the joint for about one half the depth with coal-tar paving cement and the upper part with hydraulic cement mortar. The cement mortar gives a harder wearing surface than where the entire joint is filled with pitch. It also protects the pitch from the softening action of the sun in warm weather. Where very narrow joints are employed, the greater wear of the tar cement may not be of so much importance. Some engineers fill the entire joint with coal-tar cement, while others use the cement grout alone. The object in all cases should be to make a road-surface as nearly impervious to water as possible.

The coal-tar paving cement used for the purpose of filling joints is the same as that used for brick pavements. It is usually formed of coal-tar residuum mixed with creosote oil or with tar, but it varies widely in character in different places. Sometimes asphalt is used, or asphalt mixed with coal-oil, but more frequently the name asphalt is incorrectly applied to coal-tar products, when it is called asphaltic cement. It is, of course, always applied hot to the pavement.

When the ordinary coal-tar cement filling is employed, the joints are first filled nearly full of sand or gravel, which is rounded down with a bar, after which the hot cement is poured in until the joint is well filled.

When cement grout is used it should consist of a

rich mortar, 1 part Portland cement to 2 parts sand, or 1 part natural cement to 1 part sand.

In London, where the wood blocks are set upon a layer of asphalt, it is customary to fill the lower part of the joint with melted asphaltic cement and the upper part with cement grout. This method is not, however, extensively employed.

The method of procedure in constructing a wood pavement depends upon the kind of foundation employed. When a concrete base is used, a cushion coat of sand about one inch in thickness is usually spread evenly over the concrete. Upon this layer of sand the blocks are set, close together if round blocks, or in rows as already described if rectangular blocks are used. The blocks are then rammed to a firm and even bearing by the use of heavy wooden rammers, the joints are filled with gravel and paving cement or grout, as the case may be, a layer of gravel is spread over the surface, and the pavement is opened for traffic. Most of the well-known London wood pavements are constructed in this manner, the blocks being of fir or deal, and cut to a rectangular shape, although, as already stated, some of the later ones are built with close joints from which the gravel filling is omitted.

Where the sand and board foundation is employed, the best practice is probably represented by the method pursued at Chicago, which is approximately as follows: A layer of sand 3 inches thick is placed upon the roadbed, which has been compacted by rolling. The sand is rammed or rolled until well compacted, and the foundation layer of 2-inch hemlock planks laid lengthwise of the street, resting at their ends and middles upon stringers 1 inch by 8 inches bedded firmly in the sand. Upon this foundation the cedar blocks are set close together, the joints are filled with small gravel well rammed, and the pavement is then flooded with hot tar cement so as to fill the interstices in the joints. A coating of gravel one inch thick is then placed upon the pavement, and traffic is allowed to come upon it.

In some places these pavements are constructed without the use of the coal-tar cement, the joints being rammed full of sand and gravel; in other cases the blocks are set upon rolled sand and gravel without the boards, the blocks being rammed into place; but otherwise the construction is the same as above.

The method of construction advisable for any particular work depends always upon the local conditions and requirements. To make a good wood pavement there is necessary a solid foundation, blocks of good material, and impervious joints, and all such work should be so constructed as to secure these conditions in so far as available resources will admit. Weak construction always involves high cost for maintenance, and greater expense in the end than good construction.

ART. 66. PRESERVATION OF WOOD.

The most serious objection commonly raised to the use of wood pavements is that wood, being porous, absorbs moisture readily, and is thus both liable to destruction through decay and to become injurious to health. Various methods have therefore been proposed for rendering the blocks less pervious and more durable by impregnating them with various solutions which shall fill the pores and act as preservatives.

The methods which have been principally used are known as *Burnettizing*, *Kyanizing*, and *Creosoting*.

Burnettizing consists in immersing the wood in a solution of chloride of zinc until the pores are filled with the solution. This is either done by simply immersing and allowing the wood to gradually absorb the solution, or by forcing the solution into the pores of the wood under pressure. The first method requires considerable time : about two days' immersion to each inch of thickness is usually allowed in order to admit of the wood becoming saturated with the solution.

In kyanizing a saturated solution of corosive sublimate is used, and the timber immersed in the solution long enough for the pores to become well filled.

Creosoting consists in impregnating the wood with the oil of tar or creosote. In this process the wood is first thoroughly dried, usually by heating it in a kiln, and the hot creosote is then forced in under pressure. The method of accomplishing this varies in different places. In order to be effective the process must be thoroughly carried out and the pores well filled. It is commonly recommended that from 8 to 12 pounds of creosote per cubic foot of timber should be forced in, as a minimum requirement for the softer woods, such as are commonly used in pavements. Creosote has the property of destroying the lower forms of animal life, and is therefore an effective preservative against destruction through these agencies where they exist. This method is therefore often employed for the preservation of timber for subaqueous construction in sea-water.

All of the above processes, when properly applied, are effective in preventing decay, and therefore in lengthening the natural life of the wood. They also render the wood practically impermeable, and thus remove the objection to the pavement based upon its absorbent nature. They do not, in general, appear to increase the resistance of the wood to the wear of the traffic, and in most cases the advantage to be gained seems so small as to render their economic value for this purpose at least doubtful.

The economic advantage of using treated blocks is a question of the relative costs of increasing the expense of construction by using them or of the additional expense of more frequent renewals where they would be necessary without the treatment. The desirability of the treatment in any particular instance depends to some extent upon the traffic to which the pavement is to be subjected as well as upon the character of the material available for the purpose. Where the traffic is such as to bring a considerable wear upon the pavement, and sound, well-seasoned blocks are to be had, there is usually little, if any, advantage in the treated blocks, as the pavement will ultimately fail by the wear of the blocks in either case. Experience has shown that in many cases the untreated wood wears out in the pavement before decay sets in, and that the application of the preservative processes would not prolong the life of the pavement. This has been the result of experience in London, where after trying many different methods the consensus of opinion is against the use of preservative processes. It is claimed, however, by some authorities that creosoted blocks have been shown in some instances to give greater resistance to wear than untreated blocks under the same traffic (see London *Engineering* for July 29, 1892). This London traffic is heavy, the material well selected, and the wear severe. There may frequently be cases, however, where with lighter traffic or with wood of a less durable nature or less well seasoned the application of preservatives may effect such a lengthening of the life of the pavement as to make their application economically desirable.

In the treatment of the wood it is essential that the process be very thoroughly applied in order to get good results. The process most commonly recommended is creosoting, and in order to derive any benefit from the treatment it is necessary that the pores of the wood be thoroughly filled with the oil. Merely dipping the blocks in creosote or tar, as is sometimes done, is more likely to be an injury than a benefit, and has been found in some cases to be the cause of decay by closing the pores upon the surface of the block and inducing an internal dry-rot. It is also essential to success with creosote that the blocks be thoroughly dried before injecting the creosote.

ART. 67. MAINTENANCE OF WOOD PAVEMENTS.

The ordinary maintenance of wood pavements, like that of most other pavements, consists in keeping the pavement clean and in repairing from time to time any small breaks that may appear in the surface due to imperfect material or to the settling of the foundation. These repairs would, of course, include the removal of any defective blocks and the taking up and replacing of any portion which may settle out of surface through inefficient support.

It is generally agreed that the wear of a wood surface is improved by giving it an occasional coating of small gravel, in some cases two or three times a year, and permitting it to be ground into the surface for a few days. It is an advantage also that the surface be kept sprinkled in warm weather.

When the wood pavement needs renewal or extensive repairs the surface may be relaid as with any other block pavement: if a permanent foundation be employed, by stripping the blocks from the foundation and placing a new surface in the same manner as the first one ; with a board foundation that also must be relaid.

If the pavement is cut through for any cause the surface may be replaced with the same facility as other block pavements; but where a board foundation is used it is necessary to use care in replacing in order to secure proper bond with the remainder of the foundation and prevent any subsequent settlement at the line of cut. In such cases it is also necessary to compact the earth very carefully in replacing it, that there may be no subsequent settlement.

The cost of keeping a wood pavement in order, as with any other pavement, depends upon the character of the work done in construction,—the better the pavement the cheaper the maintenance.

ART. 68. HEALTHFULNESS OF WOOD PAVEMENTS.

The use of wood pavement is very often objected to upon the ground that it is unhealthful and likely to give rise to disease. This is based upon the fact that the material of the pavement, being porous and absorptive of moisture, is likely to become saturated with organic matter from the foul liquids of the surface soaking through it. This foul matter must also, on account of the permeable nature of the material, pass to some extent through the pavement and contaminate the foundation and soil beneath as well, especially where the foundation is a permeable one, as in the case of boards or sand and gravel. In addition to the danger due to the permeability of the wood, unhealthfulness may also be caused by the liability to decay of the material.

There is much difference of opinion among authorities concerning the extent of the danger to health offered by ordinary wood-block pavements, some regarding the danger as a very serious one and protesting against the use of wood in any case, while others, although admitting the permeability and perishable nature of the material, consider its proper use quite a safe one and the danger as somewhat visionary. Health statistics of cities using wood pavements to a large extent, compared with those of cities not using them, do not indicate anything unfavorable to them; but it may properly be said that such statistics can seldom be compared in such a manner as to give a reliable index, as there are so many other circumstances which may affect public health, and the conditions other than the pavements are rarely the same.

The likelihood of a pavement producing unsanitary conditions depends very largely upon climatic and local conditions and upon the construction of the pavement. The opinions of those observing the matter are therefore usually based upon their own local surroundings. Instances are recorded where blocks of wood after considerable service in pavements have been found to be but little, if any, affected by the absorption of the street refuse, and the foundations to be quite unaffected by it; and in other instances blocks have been found to be considerably contaminated, and the foundations and subsoil saturated with filth. In most cases these differences may be attributed to differences in methods of construction as well as in material used, and the cause of subsoil contamination appears usually to be open-joint construction rather than permeable blocks.

Where good drainage exists and a pavement is constructed of sound, well-seasoned blocks with close, impervious joints, so that it cannot get wet at the base, the danger from saturation and decay is probably small, and on the score of health such a pavement is much to be preferred to a stone-block pavement with open joints. Under the reverse of these conditions a wood pavement may be a serious menace to health.

In close, damp places, or climates giving the same conditions, the liability to decay is much greater than where the pavement is exposed to the sun and air.

CHAPTER X.

STONE-BLOCK PAVEMENTS.

ART. 69. STONE FOR PAVEMENTS.

STONE-BLOCK pavements are commonly employed where the traffic is heavy and a material needed which will resist well under wear.

Stone for this purpose must possess sufficient hardness to resist the abrasive action of wheels. It must be tough, in order that it may not be broken by shocks. It should be impervious to moisture and capable of resisting the destructive agencies of the atmosphere and of weather changes.

Experience only can determine the availability of any particular stone for this use. The stone may be tested in the same manner as brick, and perhaps something predicated as to the probability of its wearing well under traffic; but the conditions of the use of the material in the pavement are quite different from those under which it may be tested, and any tests looking to a determination of its weathering properties are apt to be misleading.

Examination of a stone as to its structure, the closeness of grain, homogeneity, etc., may assist in forming an idea of its nature and value for wear. Observations of any surfaces which may have been exposed for a considerable time to the weather, either in structures or in the quarry, will be the most efficient method of forming an opinion concerning the weathering properties of the stone. The conditions of use in pavements are, however, somewhat different from ordinary exposure in structures, on account of the material in the pavement being subject to the action of water containing acids and organic substances due to excretal and refuse matter. A low degree of permeability usually indicates that a material will not be greatly affected by these influences and also that the effect of frost will not be great.

Granite and sandstones are commonly employed for paving blocks and furnish the best material. Limestones are sometimes used, but have seldom been found satisfactory. Trap-rock and the harder granites, while answering well the requirements as to durability and resistance to wear, are objectionable on account of their tendency to wear smooth and become slippery and dangerous to horses. Granite or syenite of a tough homogeneous nature is probably the best material for the construction of a durable pavement for heavy traffic. Granites of a quartzy nature are usually brittle and do not resist well under the blows of horses' feet or the impact of vehicles on a rough surface. Those containing a high percentage of felspar are likely to be affected by atmospheric agencies, while those in which mica predominates wear rapidly on account of their laminated structure.

Sandstones of a close-grained compact nature often give very satisfactory results under heavy wear. They are less hard than granite and wear more rapidly, but do not become so smooth and slippery, and commonly form a pavement that is more satisfactory from the point of view of the user. Sandstones differ very widely in character, their value depending chiefly upon the nature of the cementing material which holds them together. In order that a stone may wear well and evenly in a pavement it is desirable that it be fine-grained, dense, and homogeneous, as well as cemented by a material which is not brittle and is nearly impervious to moisture. Those sandstones in which the cementing material is of an argillaceous or calcareous nature are apt to be perishable when exposed to the weather. The Medina sandstones of Western New York and Ohio have been quite extensively used for paving purposes and prove a very satisfactory material for such use.

Limestone has not usually been successful in use for the construction of block pavements on account of its lack of durability against atmospheric influences. The action of frost commonly causes weakness and shivering, which produces uneven and destructive wear under traffic. There are, however, as wide variations in the characteristics of limestones as in those of sandstones, and there may be possible exceptions to the rule that in general limestone is not a desirable material for block pavement

ART. 70. COBBLESTONE PAVEMENTS.

Cobblestones have in the past been quite extensively used in the construction of street pavements, although at the present time they have been for the most part abandoned, excepting where they are used at the sides of other pavements for gutter construction or sometimes between the rails of a horse-car track. This pavement as ordinarily constructed is a cheap one in first cost, and it affords a good foothold for horses. It is not usually a durable pavement as the stones are easily loosened from their positions, although the stones themselves may be practically indestructible and used again and again in reconstructing the surface.

Cobblestone pavements as commonly constructed are also objectionable because they are permeable to water and difficult to clean. They therefore collect and become saturated with the filth of the street and are very liable to injury from frost. They are also extremely rough and unsatisfactory in use for travel.

For paving the side-gutters, where broken-stone or sometimes where wood is used for the travelled portion of the street, cobblestones may often be convenient and useful, and form a cheap and satisfactory means of disposing of surface drainage. Such an arrangement is shown in Fig. 31 (p. 186).

Cobble pavements may also sometimes be advantageously used upon steep grades where traffic is necessarily slow and the foothold afforded becomes a very important matter. When used for this purpose a concrete foundation should be employed and the stones be firmly bedded to prevent displacement through the efforts of horses to obtain foothold.

Cobblestones as used for pavements are usually rounded pebbles from 3 to 8 inches in diameter. They are set on end in a layer of sand or gravel, rammed into place until firmly held in position, and then covered with sand or fine gravel and left to the action of travel, which soon works the upper layer of sand into the interstices between the stones.

Art. 71. Belgian Blocks.

Belgian block is the name commonly applied to a pavement formed of nearly cubical blocks of hard rock. In the vicinity of New York this pavement has been largely used, the material being trap-rock from the valley of the lower Hudson. The blocks are usually from 5 to 7 inches upon the edges, with nearly parallel faces, and as commonly laid are placed upon a foundation layer of sand or gravel about 6 inches thick. This shape of block is objectionable on account of the width between joints being too great to afford good foothold to horses. The materials of which Belgian blocks have ordinarily been formed are very hard and (as already noted in Art. 69) wear smooth in service, becoming slippery and thus increasing the effect of the too wide block. It is also better to have the length of the blocks somewhat greater across the street and let them break joints in that direction in order that they may give greater resistance to displacement under passing wheel-loads.

The older pavements of this character were usually placed upon a sand foundation. More recently, this practice has, in the better class of work, been superseded by a more solid construction, a concrete base being used.

Art. 72. Granite and Sandstone Blocks.

For the construction of the better class of stone-block pavements, blocks of tough granite or sandstone are used, set, in the best work, upon a concrete base, although sometimes placed upon a foundation of sand or gravel.

These pavements when well constructed are about the most satisfactory means yet devised for providing for very heavy traffic, as they present a maximum resistance to wear with a fairly good foothold for horses, and are much more agreeable in service than the old form of rough pavements. There is still much to be desired in the attainment of smoothness and absence of noise, and, as a general thing, it may be said that pavements of this kind are desirable only where the weight of traffic is so great that the smoother pavements would not offer sufficient resistance to wear. Even in such cases it may frequently be questionable whether an additional expense for maintaining a pavement which would be more pleasant in use and less objectionable to occupants of adjoining premises would not be advisable from an economical as well as from an æsthetic point of view.

Blocks for stone pavements, in the best work, are cut in the form of parallelopipeds, 9 to 12 inches long, 3 inches wide, and 6 or 7 inches deep. The length should be sufficient to permit the blocks to break joints across the street. The width should be less than that of a horse's hoof in order that the joints in the direction of travel may be close enough together to prevent a horse from slipping in getting a foothold. The depth should be sufficient to give a bearing surface in the joints large enough to prevent the blocks from tipping when the load comes upon one end of it.

ART. 73. CONSTRUCTION OF STONE-BLOCK PAVEMENTS.

Stone-block pavement for durable and effective service should be placed upon very firm foundations.

Bases of concrete are usually employed and give the best results. These foundations are formed as described in Art. 47, and consist of a layer of concrete 4 to 8 inches thick, 6 inches being the most common depth.

In constructing the pavement, a cushion coat of sand, usually about an inch thick, is spread upon the base of concrete, for the purpose of allowing the bases of the paving blocks to be firmly bedded when the tops are brought to an even surface, the sand readily adjusting itself so as to fill all the spaces beneath the blocks and to offer a uniform resistance to downward motion in every part of the pavement, and in like manner transmitting the loads which come upon the pavement to the foundation so as to evenly distribute them over the surface of the concrete. The sand used for this purpose should be clean and dry, and all large particles sifted out, as they may prevent the blocks adjusting themselves properly. A thin layer of asphaltic cement is sometimes used in place of the sand with very good results.

The blocks should be laid as close together as possible in order to make the joints small. They are laid, like brick, with the longest dimension across the street, and arranged in courses transverse to the street, with the stone in consecutive courses breaking joints. In laying, it is considered best to begin the courses at the gutters and work toward the middle, the crown-stone being required to fit in tight.

After the blocks are placed they are well rammed to a firm unyielding bearing and an even surface. Stones that sink too low under the ramming must be taken out and raised by putting more sand underneath.

As in the case of other block pavements, those of

stone should be made as impervious to moisture as possible. The foundation should be kept dry and moisture prevented from penetrating beneath the blocks, where it has a tendency to cause unequal settlement under loads, or disruptions under the action of frost. In the better class of work, therefore, the joints are filled with an impervious material which cements the blocks together. Coal-tar paving cement is commonly employed for this purpose, as with brick and wood, and seems the most satisfactory in use, although hydraulic cement mortar is sometimes used. The coal-tar cement is commonly made by mixing coal-tar pitch with gas-tar and oil of creosote, a proportion sometimes employed being 100 pounds pitch, 4 galls. tar, and 1 gall. creosote.

The use of cement between the blocks binds them together and increases the strength of the pavement as well as the resistance of the blocks to being forced out of surface. It also deadens to some extent the noise from the passing of vehicles where asphaltic or coal-tar cement is used.

The method of filling the joints is usually to first fill them about one third full of small gravel, then pour in the paving cement until it stands above the gravel; then another third full of gravel, more cement as before; then gravel to a little below the top, and the joint filled full of cement; after which a coating of fine gravel is distributed over the surface.

Various modifications of the method above outlined are used in the principal cities for a pavement to withstand heaviest traffic and secure a maximum of durability: essentially it represents the best modern practice.

A cheaper form of stone-block pavement is made by

laying the blocks directly upon a foundation of gravel or sand, either with cemented joints or with joints filled with gravel only. This gives a fairly good pavement for streets of moderate traffic, and has been extensively used in the past. The present tendency, however, which will probably increase in the future, is to lessen the use of pavements of this character, and to substitute a surface which is more pleasant in use for all service where durability and resistance to wear are not the prime requisites.

ART. 74. STONE TRACKWAYS.

In some of the European cities, particularly in Italy, stone trackways are sometimes employed on streets of heavy traffic for the purpose of diminishing traction. These trackways are formed of smooth blocks of stone 4 to 6 feet long, 18 to 24 inches wide, and 6 to 8 inches deep, laid flat and end to end so as to form a smooth surface upon which wheels may move with the least possible resistance. Between the tracks, and usually the remainder of the street, is commonly paved with cobble. The method of construction is shown in Fig. 25. The tracks drain to the middle, and the pavement

FIG. 25.

between is made concave and provided with openings into the storm sewers for the escape of surface-water. The track and pavement are laid upon a layer of sand resting upon a broken-stone or gravel foundation.

Such trackways are very durable under heavy traffic, and give very light traction combined with good foothold. It is possible that they might advantageously be applied oftener than they are on streets used for heavy hauling.

CHAPTER XI.

CITY STREETS.

ART. 75. ARRANGEMENT OF CITY STREETS.

THE location of streets should be planned with a view to giving direct and easy communication between all parts of a city. The arrangement should also be such as to permit the subdivision of the area traversed by them in such a manner as to give the maximum of efficiency for business or residential purposes. The most obvious and satisfactory method of accomplishing these purposes is usually by the use of the rectangular system, with occasional diagonal streets along lines likely to be in the direction of considerable travel.

Streets so far as possible should be systematically arranged and continuous throughout the extent of the city, both to facilitate travel and to admit of their being so named and numbered that the locality of a place of business or residence may at once be evident, from its address, to any one familiar with the general plan of the city. The rectangular system is desirable on this account, and also because it furnishes blocks of the best form for subdivision into building lots.

The proper arrangement of streets will always necessarily depend in some measure upon the natural features of the locality, and any system of arrangement will be more or less modified by local topography.

Where for topographic or æsthetic reasons it may be considered desirable to use curved lines for the streets, the continuity and uniformity of arrangement should be maintained as far as possible. The use of curves on residence streets may sometimes be advantageous in reducing gradients or in its effect upon adjoining property through avoiding heavy earthwork. Where a change in direction is necessary the use of a curve usually gives a better appearance than an abrupt bend, unless the change can be effected at the intersection of a cross-street. Care is required, however, to prevent the local introduction of curvature disarranging the general plans and producing the chaotic condition due to an irregular use of short streets.

In laying out a rectangular system of streets the blocks ordinarily will preferably be long and narrow. The distance needed between streets in one direction is only that necessary to the proper depth of lots, while in the other direction the streets need only be close enough to provide convenient communication for the travel and traffic. A convenient method would be to lay out the main streets so as to form squares large enough to permit the introduction of an intermediate minor street through the blocks. These minor streets may then be introduced in the direction that seems advisable in each locality. Such an arrangement is shown in Fig. 26. The diagonal streets cut more space from the blocks traversed by them, but give more frontage, and property fronting them will usually have more value than other property in its vicinity.

The proper location for diagonal streets intended as thoroughfares for traffic is naturally determined by the positions of the business centres or public buildings and parks, from which they may radiate in such manner as to bring the outlying portions of the city into the most direct communication possible.

A city cannot usually be laid out complete. Its formation is a matter of gradual growth and enlargement,

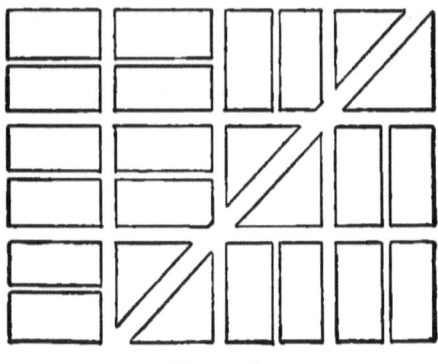

FIG. 26.

and the end cannot be seen from the beginning. For this reason it is frequently necessary to undergo great expense in the larger cities in cutting new streets or in changing the positions or dimensions of existing old ones in built-up districts in order to relieve the crowded condition of the streets, which hampers business and renders travel difficult and unpleasant. Much of this difficulty might frequently be obviated if in growing towns and cities proper attention were given to the regulation of suburban development. Such development should be under municipal control so far as to require at least that each new subdivision which opens new streets should be made with a view to affording proper ways of communication between adjoining properties by making streets continuous. Where such regulation does not exist streets will be

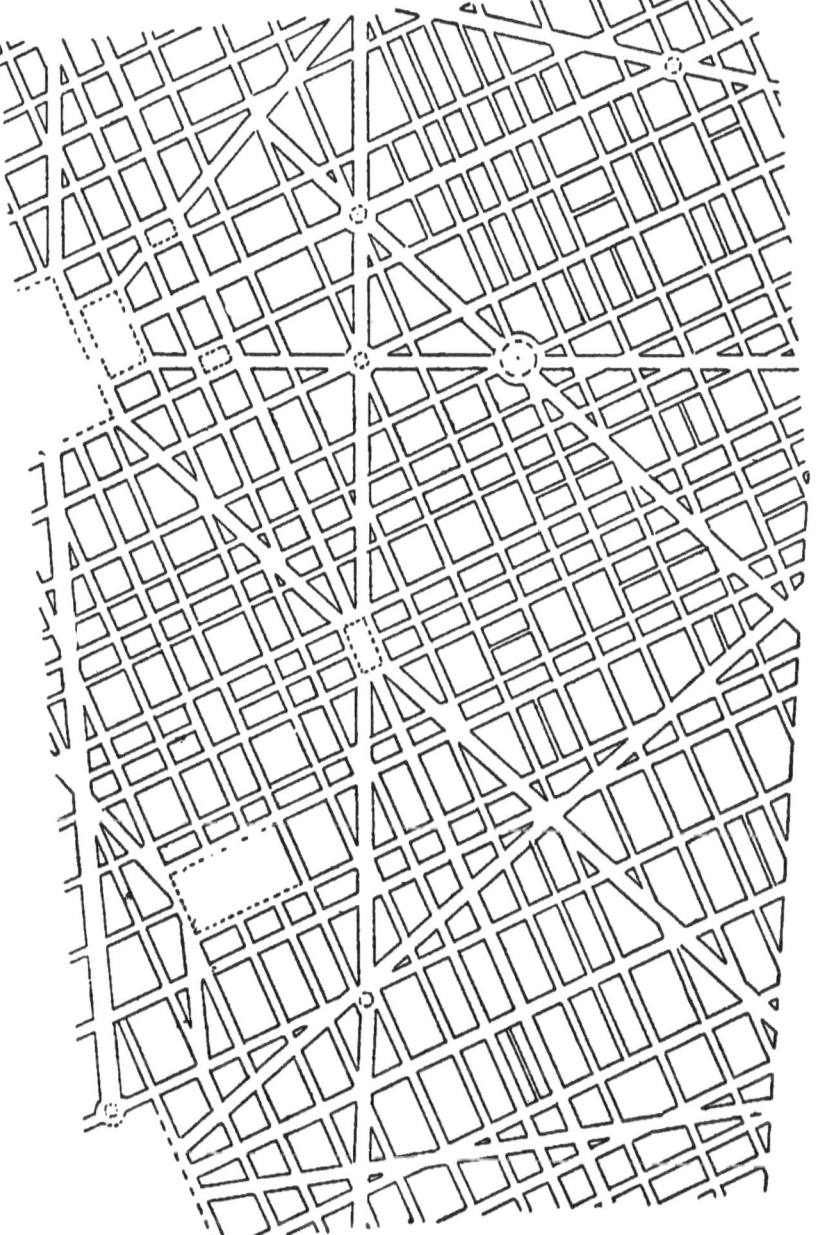

Fig. 27.

182   A TEXT-BOOK ON ROADS AND PAVEMENTS.

laid in any manner to best develop the particular property in which they are placed.

A good example of the advantages of systematic and liberal plans in street arrangement, as well as of the evils of unregulated extension, is given by the case of Washington, D. C.

Fig. 27 shows a portion of the city of Washington illustrating its systematic arrangement. It consists of a rectangular system, together with two sets of diagonal avenues, and open squares or circles at the intersections of the avenues.

Fig. 28 shows a number of suburban subdivisions on the borders of the city of Washington, made previous to the adoption of the law regulating them. In some

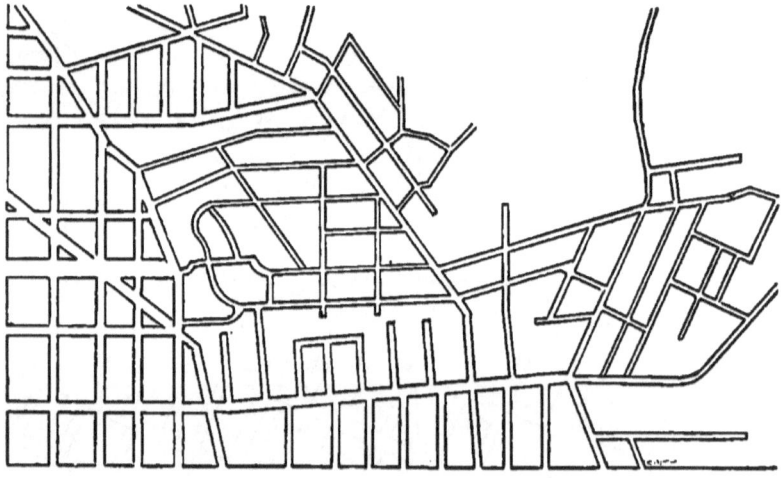

FIG. 28.

cases the streets of adjoining subdivisions have no communication with each other, and the general tendency is toward a labyrinth of short streets. The law now requires that all street extension within the District of Columbia shall conform to the general plan of the city of Washington; and under the operation of this law the lines of many of the city streets have been extended to all parts of the District, and all of the suburban development is being gradually brought with the city into one harmonious whole, on the same generous plan that exists within the city. The rectification of the irregular plats upon the borders of the city must, however, be a matter of heavy expense to the District.

ART. 76. WIDTH AND CROSS-SECTION.

The width of city streets is important both on account of its influence upon the ease with which traffic may be conducted, and because of its effect upon the health and comfort of the people, by determining the amount of light and air which may penetrate into thickly built-up districts.

To properly accommodate the traffic of commercial thoroughfares in business districts of towns of considerable size, a street should have a width of 100 to 160 feet, the whole of it to be used for roadway and sidewalks. Wide streets are especially needed where, as in the larger cities, they are bordered by high buildings or are to carry lines of street railway.

Residence streets in a town of considerable size, where houses are set out to the property line and stand close together, should have a width of at least 80 to 100 feet in order to look well and give plenty of light and air.

The streets in nearly all large towns are laid out too narrow; they are crowded and dingy. The chief difficulty is that the future of a street is not usually foreseen when it is located. Owners in subdividing property are only anxious to get as many lots as possible out of it, and there are usually no regulations looking to the future health and comfort of resident when the street shall be built upon. In the growth of a town the nature of localities change: residence streets become business streets, streets devoted to retail trade become wholesale streets, and mercantile districts are given up to manufacturing. If a city could be laid out complete from the beginning it would be comparatively easy to consider the requirements to be met and locate the streets accordingly. Under existing conditions this is not possible, but a more liberal policy in planning streets would usually be found of advantage in any growth that may ensue. There is also very frequently an immediate financial advantage in the enhancement of values due to wide streets. A lot 100 feet deep on a street 80 feet wide will nearly always be of greater value than if the same lot be 110 feet deep and the street only 60 feet in width.

In Washington, D. C., which probably has the best general system of any American city, no new street can be located less than 90 feet in width, and avenues must be at least 120 feet wide. Intermediate streets, called places, 60 feet wide, are allowed within blocks, but full-width streets must be located not more than 600 feet apart. The value of this liberal policy to the city of Washington is evident not only in the increased comfort of the people, but in its large growth as a residential city and the increased value of property in it.

While it is advantageous to have the street wide between building-lines, it is not necessary that the whole street width be used for pavements. The street pavement should be gauged in width by the immediate necessities of the traffic which is to pass over it. The pavement should be wide enough to easily accommodate the traffic, but any unnecessary width is a tax upon the community in the construction and maintenance of more pavement than should be required, and perhaps diminishes the length of street which may be improved with available funds. Thus, for a residence street in general a width of 30 to 35 feet between curbs is usually ample, with a foot-walk upon each side 6 to 10 feet wide. The remainder of the street width should be made into lawns upon each side, with tree spaces between the sidewalk and roadway.

Fig. 29 shows in partial section the arrangement of a

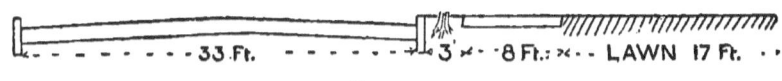
- - - - - - - 33 Ft. - - - - - - - 3' - - 8 Ft. - - LAWN 17 Ft. - -

FIG. 29.

90-ft. residence street for moderate traffic. For residence streets of lesser importance, where the travel is light and the street is only required to furnish facilities to meet the needs of its immediate locality, a less width of pavement may often be advantageously used. A pavement 24 feet wide is sufficient to accommodate a very considerable amount of light driving, and in many places, especially in the smaller towns where funds for effective improvement are obtained with difficulty, even less widths may be employed with the result of improving the streets both in appearance and usefulness. All that is really needed in such cases is room for teams to pass comfortably and to turn without difficulty. The narrowing of roadways on streets of light traffic to what is really necessary may often make possible improvements which will turn a broad sea of mud into a narrow hard roadway and a grass-plat. Fig. 30 shows the arrangement of a village street 50 feet wide for light service.

In many cases for village streets, where the traffic is light and it is essential that the cost of construction be

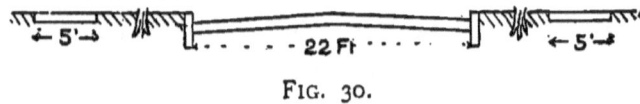

FIG. 30.

low, it may be good practice to construct the travelled portion of the roadway of macadam, wood, or other pavement, and use cobble gutters at the sides without curbs. Fig. 31 shows a roadway 30 feet wide, with macadam middle and cobble gutters. In Saginaw, Mich., this method has been followed, using either macadam or wood blocks for the middle portion, and

FIG. 31.

in the report of City Engineer Roberts for 1893 it is recommended as economical and efficient.

The cross-section of streets must be arranged with reference to proper surface drainage. The street is given a crown at the middle to throw the water into the gutters, and sidewalks usually have a sufficient inclination toward the gutter to cause them to drain over the curb. The section necessary for street drainage is discussed in Art. 10. The street is usually made practically level across, the curbs and sidewalks at the two sides being given the same elevation. The parking at the sides may have a slope between the sidewalk and the building-line when it is necessary or advantageous. Sometimes, on streets along a slope, expense may be saved or adjoining property benefited by placing the

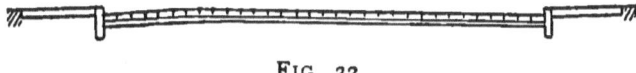

FIG. 32.

sidewalk at a different elevation from that of the street, as shown in Fig. 5, or by placing one curb lower than the other and moving the crown of the road to one side, as shown in Fig. 32.

ART. 77. STREET GRADES.

The grades of city streets necessarily depend mainly upon the topography of the site. Wherever possible, it is desirable that grades be uniform between cross-streets.

In establishing grades for new streets through unimproved property, they may usually be laid with reference only to obtaining the most desirable gradients for the street within a proper limit of cost. But where improvements have already been made, and located with reference to the natural surface of the ground, it is frequently a matter of extreme difficulty to give a desirable grade to the streets without injury to adjoining properties. In such cases it becomes a question of how far individual interests shall be sacrificed to the general good. It may be said in this connection that adjustments to new grades are usually accomplished much more easily than would be anticipated, and when accomplished the possession of a desirable grade is of very considerable value to adjoining property. Too great timidity should not, therefore, be felt in regard to making necessary changes because of the fear of injuring property in the locality.

Where a grade if made continuous between intersecting streets would be nearly level, it is frequently necessary to put a summit in the middle of the block and give a light gradient downward in each direction to the cross-streets in order to provide for surface drainage. The amount of slope necessary to provide for proper drainage depends upon the character of the surface and smoothness of the gutter. For a surface of earth or macadam the slope should not be less than about 1 in 100, and for paved streets from 1 in 200 to 1 in 250.

In some cases it may be possible to give sufficient slope to gutters to carry off the surface-water by making the gutter deeper at the ends than in the middle of the block without making a summit in the crown of the street. The curb in such case would be made level or of uniform gradient.

The smoother forms of pavement are only applicable to light gradients. Rock asphalt is usually limited to 2 or $2\frac{1}{2}$ per cent grades. Trinidad asphalt may be used to grades of about 4 per cent. Brick, if kept clean, is safe on gradients of about 6 per cent; wood, on those a little steeper; and stone blocks are satisfactory to about a 10-per-cent gradient.

Pavements on steeper gradients must be made rough in order to insure a safe foothold to horses. On grades steeper than 9 or 10 per cent cobblestones are preferable to rectangular stone blocks, as they give better foothold, and the speed of travel being necessarily slow the roughness is of less consequence. For such use it is desirable to have the cobblestones set on a concrete foundation and the joints filled with paving cement after the manner of a first-class block pavement, as the wear on a steep slope will be severe. Ordinary stone blocks may be laid on steep streets with wide joints, about an inch, so as to give better foothold than the common form; or the corners of the stones may be bevelled on the upper edges and set in the usual manner.

In a report on the streets of Duluth in 1890, Messrs. Rudolph Hering and Andrew Rosewater recommend for steep streets, in addition to the above, that brick may be used in which the tops are rounded, and that wood blocks for such use have their upper edges chamfered on each side, or if round blocks be used, around the blocks.

ART. 78. STREET INTERSECTIONS.

At intersections the crown of the roadway pavement on each street should, if possible, be continuous to the centre of intersection, in order to prevent vehicles on one street from being subjected to the jar incident to passing over the gutter of the other. Where a storm sewer is available into which the water from the gutters on the upper side can be emptied this is a simple matter, but where such sewers do not exist it requires the adoption of some special means of draining the gutters on the upper side. This may sometimes be accomplished by a culvert across the street, the gutters being somewhat depressed at the corners to bring the channel sufficiently low. In other cases, where the slope is sufficient, it is more satisfactory to construct an underground pipe-drain from the upper corner to some point in the gutter below the crossing.

Where the rate of grade is such that it is feasible, it is desirable that the grade of both streets should be brought to a level at intersections. The top of the curb at the four corners should be at the same elevation, thus permitting the continuation of the full section of each roadway until they intersect. It is also desirable that the sidewalks at the corners be level; that is, the points aa in Fig. 33 should all be placed at the same eleva-

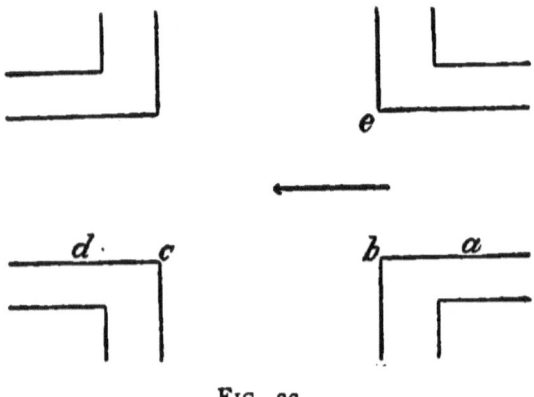

FIG. 33.

tion, which will make the entire street section, including sidewalks, horizontal across the direction of travel on each street.

On very steep slopes it may not be possible to flatten out the grade to a level in crossing transverse streets, and in such cases the elevations require study, and need to be carefully worked out for each particular case. In the report of Messrs. Rudolph Hering and Andrew Rosewater upon the streets of Duluth, it is recommended that in all cases the grade shall be reduced to 3 per cent between the curb-lines of cross-streets, and the grade of the curb reduced in all cases to 8 per cent for the width of the sidewalks of intersecting streets. This is to be considered the maximum allowable rate of transverse grade, and only to be employed in case of necessity. If in Fig. 33 the arrow represents the direction of steep slope, and the street transverse to that direction has a roadway 40 feet wide with sidewalks 10 feet wide, the above limits would permit the curb at c to be 1.2 feet lower than that at b, and admit of a fall of 0.8 foot in the curb line from a to b and from c to d. If both streets have the same grade and width the curb at the lowest corner would be 2.4 feet lower than at the highest corner.

Sometimes, where the parallel streets in one direction follow the lines of greatest slope, and the cross-street are normal to them, the proper grades at intersections may be arranged by giving the streets along the slope a section similar to that shown in Fig. 32 throughout its length, thus permitting the street in the direction of slope to continue its grade across the intersection without altering at that point the side slope of the cross-street.

For a case of maximum slope this would make the section of the roadway of the cross-street a plane surface sloping uniformly from the upper to the lower curb, or in Fig. 32 it would transfer the street crown to the upper curb.

Art. 79. Footways.

Footways are not required to bear the heavy loads which come upon the roadway pavement, but in streets of considerable travel are subjected to a continual abrading action, and for good service are required to be of a material which will resist abrasion well, of so uniform a texture as to wear evenly, and not hard enough to become smooth and slippery in use.

A good sidewalk should always present an even surface, and therefore requires a firm foundation to resist the displacement of the blocks of which it may be composed. It must also be durable under atmospheric changes, and of material that may be easily cleaned. The materials commonly employed are gravel, wood, brick, tar, asphalt, stone, and artificial stone.

Gravel walks are the cheapest of footways where suitable material is available. They are constructed in a manner similar to that used for gravel roadways, and require that the bed of the walk be well drained, and that it be well compacted by rolling or ramming before the walk is placed upon it. The best gravel walks are usually built upon a base of rough stone. This base may be 6 or 8 inches thick, and forms a solid foundation upon which the gravel surface may be placed and sustained against settling. Walks constructed in this manner are frequently used in city parks where the travel is considerable. On suburban roads, gravel walks usually consist of a thin surface of gravel laid upon the earth-bed, and are replaced by some other surface when a more expensive construction can be afforded. Gutters are frequently necessary to protect the walks from the wash of surface-water, which otherwise very quickly destroys it.

Wood is commonly used for walks in the form of planks, which are laid on stringers, the planks being placed perpendicularly to the direction of travel. It is comparatively short-lived, and requires considerable expenditure for repairs, as the material is perishable and also wears rapidly.

Brick footway pavements have been extensively used for many years, and form, when well constructed, a very durable and satisfactory sidewalk. As commonly constructed they consist of ordinary hard-burned bricks laid flat upon a layer of sand over the earth-bed. For light travel, pavements so constructed may last well and give good service; but they are apt to soon become uneven through the sinking of the bricks because of insufficient foundation.

In constructing such a pavement the sand layer should be well compacted by rolling or ramming before setting the bricks, which should also be rammed to a firm and even bearing. To give satisfactory results, a foundation of sand and gravel or broken stone should be formed 8 or 10 inches in thickness. In Washington a layer of gravel 4 inches thick and well compacted is used, with a layer of sand of the same thickness upon it to receive the surface. In forming the pavements, the bricks are laid flat and as close as possible. The joints are filled with sand, usually by coating the surface with a layer of sand before ramming, and after completion a second coating, which is allowed to remain a few days after admitting the travel to it.

Care must be used in selecting brick for this purpose to get only hard-burned brick of uniform quality, in order that the resistance to wear may be even. The use of vitrified paving brick, as used for roadway pavement, would be of advantage on walks subjected to heavy wear.

The use of a concrete foundation and setting the

brick on edge and in mortar, after the manner of constructing a roadway pavement, makes a very durable sidewalk under heavy travel. It is, however, somewhat expensive, and usually a stone surface would be preferable where such expense is to be incurred.

Footway pavements of a concrete in which coal-tar is the binding material have been widely used, but have not usually been satisfactory in use. As commonly constructed they wear rapidly and soften, becoming very disagreeable in hot weather. Some pavements of this character have, however, shown fairly good service.

Numerous methods have been proposed and tried for the construction of tar foot-walks, differing from each other in the materials mixed with the tar to form the concrete, and in the manipulation of the process. Ashes mixed with sand and gravel are usually employed, and sometimes clinkers from an iron foundry. A somewhat successful pavement of this class has a small amount of Portland cement mixed with the ashes and sand used in forming the concrete before the addition of the tar.

Asphalt footway pavements are formed either of asphalt blocks or of a surface of sheet asphalt. Where blocks are used they are laid in the same manner as brick upon a foundation of sand or gravel. The blocks, or tiles as they are commonly called, are usually made flat, about 8 inches square and 2 to $2\frac{1}{2}$ inches thick. They are laid with their edges either at right angles to the street line or at an angle of 45° with the street line,—usually at right angles, on account of greater ease in laying.

Sheet-asphalt footways are laid in the same manner as an asphalt street pavement, the pavement, however, being given a less thickness. In Washington, D. C., these pavements are made about 3 inches thick, and constructed upon a bituminous base. Material removed from street pavements in resurfacing is used for forming the surface material of the footway. Mixtures of coal-tar and asphalt similar to that used for distillate pavement, as noted in Art. 61, are also used in footways, and are commonly spoken of as asphalt.

In the use of rock asphalt for footways, the asphalt mastic mentioned in Art. 57 is commonly used, mixed with sand or gravel to give a wearing-surface. The ingredients are heated together and applied hot to a broken-stone or concrete foundation. In Europe hydraulic cement concrete is used for the base, as in the driveways. A layer of 3 or 4 inches of concrete is employed, with a surface layer of rock asphalt or asphalt mastic and sand, $\frac{1}{4}$ to $\frac{3}{4}$ inch in thickness, for ordinary work.

Natural stone for foot-walks is ordinarily used in the form of flagging. Where flagstones of proper size and good wearing qualities may be readily obtained, this kind of pavement, if well laid, makes a durable and satisfactory foot-walk. Flagstones should be set upon a solid foundation and be firmly bedded so as to preserve an even surface. They should not be laid, as is common in many places, directly upon an earth-bed, but should have a cushion layer of sand or of some porous material to prevent unequal settling under the action of frost.

Artificial-stone pavements, when well constructed of good materials, make the most satisfactory of footways. They form an even surface, quite agreeable in service, and are durable and economical where exposed to considerable travel. Pavements of this kind are either constructed of blocks of material made at a factory and carried to the site of the walk, or the stone is formed in the position in which it is to be used. The latter plan is more commonly followed and admits of the use of larger blocks, the size in this case being only limited by the necessity of providing for changes of dimension with those of temperature, very large blocks being liable to crack under such changes.

There are a number of methods of preparing artificial stone for pavements, many of them patented, differing to some extent in the composition of the material or the details of the work. In general the process consists in placing a layer of concrete 4 to 6 inches thick upon a layer of gravel or other porous material. A surface of rich mortar or concrete, composed of hydraulic cement with sand or crushed granite, is given to the pavement, and the surface is commonly roughened by scratching lines upon it before it is hardened. As with all other concrete-work, the pavement needs to be kept damp and protected from the sun until the mortar is fully set. A layer of damp sand spread over the surface may be advantageously employed to protect it for several days after it is opened to travel.

Art. 80. Curbs and Gutters.

Curbs are usually set in the streets of towns at the sides of roadway pavements for the purpose of sustaining and protecting the sidewalk or tree space, and of forming the side of the gutter. They are commonly formed of natural stone, but sometimes also of artificial stone, clay blocks, or cast iron.

The curbs used in different places vary considerably in form and dimensions. Stone curbs vary from 4 to 12 inches in width and from 8 to 24 inches in depth. They are usually employed from 3 to 6 feet in length, and set with close joints.

The depth must be sufficient to admit of their being firmly bedded, and to prevent overturning into the gutter. The front of the curb should be hammer-dressed to a depth greater than its exposure above the gutter, and the back deep enough to permit the sidewalk pavement to fit close against it where the sidewalk adjoins the curb. The ends of the blocks should also be dressed to the depth of exposure, and the part below the ground trimmed off so as to permit the dressed ends to come in contact when laid.

Granite is usually considered the best material for curbs, although both sandstones and limestones are used in many places. In the vicinity of New York the North River bluestone has proved a good material for the purpose.

There are various ways of setting the curb. The object should be to bed it firmly on a solid foundation. The best method is to place a bed of concrete under it. This construction is shown in Fig. 34, which represents the method used in setting granite curb in Washington, D. C. The curb is held firmly in place by the concrete foundation, which joins it rigidly to the roadway pavement.

Where the concrete foundation is not used under the curb a deeper curbstone is necessary, usually from 18 to 24 inches in good work. Curbs are very commonly set in the natural ground, the pavement coming against it on one side; but it is usually found advantageous to lay them upon a bed of gravel or broken stone, with gravel filled in the trench about them.

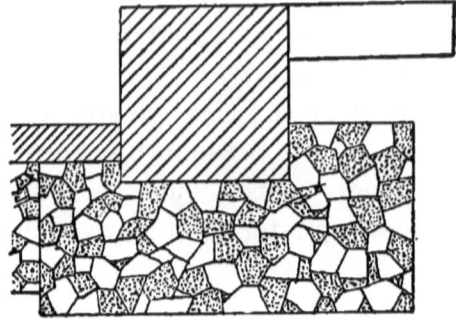

FIG. 34.

The ordinary method of setting curbs is shown in Fig. 35.

The Washington specifications for ordinary work require that a bed of gravel 4 inches deep be used under the curb, and that the trench be filled with

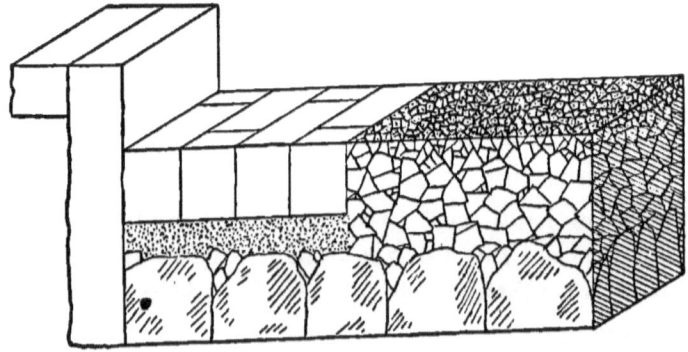

FIG. 35.

gravel placed in layers 3 or 4 inches deep, each layer being thoroughly rammed before adding the next.

Curbs of artificial stone or concrete are usually

formed by mixing the concrete upon the ground and placing it in the position it is to occupy, using a board mould, as in constructing artificial-stone foot-walks, to give it proper shape. By this method of construction the curb and gutter may be made practically in one piece, where a concrete base is used for the pavement. The concrete for the curb and gutter is made of smaller materials and with a higher percentage of cement than in preparing the foundation for the roadway, and is given a surface coating of cement mortar which is commonly formed of a mixture of Portland cement with finely crushed granite.

Specifications for artificial-stone curb in Washington, D. C., require that the concrete be composed of 1 part Portland cement, 2 parts clean sharp sand, and 3 parts clean broken stone not more than 1 inch in their largest dimensions. The exposed surface of both gutter and curb is to be coated $1\frac{1}{2}$ inches thick with a mortar composed of 3 parts granulated granite (the fragments being of such size as to pass through a $\frac{1}{4}$-inch screen, and free from dust) and 2 parts cement.

Artificial-stone curbs are sometimes made hollow, and the interior spaces used as a conduit for pipes or wires. A variety of forms are used for these cases, the curb being usually made in blocks at a factory and set like natural stone, the blocks being commonly formed in separate parts which may·be fitted together to form the curb and removed to give access to the openings. Where the hollow curbs are in one piece, hand-holes are placed at short intervals to admit of using the openings; this may be done in case the conduits are to be used for wires.

Curbs of burned clay or brick are made in several

forms, both solid and hollow, and are frequently used on streets paved with brick, where stone suitable for curbing is lacking.

Cast-iron curbs are sometimes employed, although they have not come into use extensively. They consist usually of a casting similar to that shown in section in Fig. 36, which forms the face and top of the curb, being open at the back and braced with ribs at short intervals of length. They are held in place by ties attached to the ribs, and the backs are filled and tamped to a firm bearing.

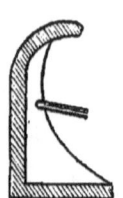

FIG. 36.

Wrought-iron plates or angles are sometimes used as a protection to concrete, or to resurface a worn stone curb, the iron being fitted to the face of the curb so as to form the exposed surface. Several forms are used, and the process is patented.

Gutters are commonly formed of the same material as the roadway pavement, which is simply extended to the curb.

In streets paved with brick or granite blocks the gutter blocks are sometimes turned lengthwise of the street, as shown in Fig. 22, for the purpose of facilitating the flow of water in the gutter. As already pointed out, however, this has the effect of making a continuous joint between the pavement and gutter, and its utility seems doubtful.

For streets paved with broken stone it is common to employ stone gutters, formed of cobblestones, of narrow flags laid lengthwise of the gutter, or sometimes of rectangular blocks. Such construction is shown in Fig. 35. On streets paved with wood these gutters may also be frequently employed with advantage, especially where for any reason the gutter is likely to be kept damp. In forming a cobble gutter the stones are usually set upon a layer of sand or gravel after the manner of forming a cobble pavement. They should be firmly bedded and form an even surface.

Cobble gutters are often used on village streets where no curbs are set, and in such locations where but slight expense is admissible they are quite satisfactory if properly constructed. This method of construction is illustrated in Fig. 31.

Sometimes in work of this kind a flagstone is used for the bottom of the gutter and the sides are formed of cobble. This is preferable as affording a more free channel for the flow of the surface drainage.

To obtain satisfactory results it is always necessary that the foundation be of sufficient depth and well compacted, in order to prevent the surface becoming uneven by the stones being forced downward into the road-bed in wet weather or through the action of frost. A layer of 6 to 10 inches of gravel or sand is usually required.

Where flagstones are used to form the gutter, they should be 3 or 4 inches thick, 10 to 15 inches wide, as may be required, and about 3 feet long. Care is required in laying that they may have an even bed and be well supported by the foundation.

Gutters of bricks, or of stone blocks, are often used for streets upon which the roadway pavement is asphalt, on account of the liability of the asphalt being injured by dampness. In this case the gutter is constructed by setting the bricks or blocks with their greatest length along the street. They are placed upon a bed of concrete, the same as is used for the foundation under the asphalt surface, and the joints are filled with coal-tar paving cement, as in constructing brick pavement.

It is also advisable in using flagstone gutters that consecutive blocks should have different widths, differing by 2 or 3 inches, in order that there may not be a continuous joint between the flagstones and the pavement of the travelled roadway.

ART. 81. CROSSINGS.

On streets paved with a smooth hard surface which is easily cleaned, such as brick or asphalt, special footway crossings are not usually required or desirable, unless the foot travel be very considerable. On other pavements, however, which are apt to be rough to walk upon or muddy in bad weather, as upon stone, wood, or macadam, footways of flagstones are commonly provided, and form the most satisfactory crossing.

These crossings consist of flagstones about 10 or 12 inches wide laid in rows across the street, the rows being 6 or 8 inches apart and paved between with stone blocks set in the ordinary manner. The crossing-stones are 3 or 4 feet long, and at least 6 inches thick in order that they may not be broken by the traffic. They should be laid with close joints and firmly bedded upon the foundation.

At street intersections where the number of pedestrians is large it is desirable that the crossing be carried across on the level of the top of the curb without leaving a step at the gutter crossing. This may be accomplished by bridging over the gutter with

CITY STREETS. 203

a flagstone or iron plate, or by placing the outlets for surface drainage a few feet back from the corner and eliminating the gutter at the corner.

ART. 82. STREET-RAILWAY TRACK.

Track for street railways upon paved streets should be constructed with a view to offering as little obstruction to ordinary street traffic as possible, while permitting the ready operation of the railway. These two points are apt to conflict, and the interest of the railway company in the construction of track is rarely identical with that of the public use of the street.

Track in streets is usually constructed of rails laid upon cross-ties, either fastened directly to the ties as in the track of steam roads, or supported upon chairs which serve to raise the surface of the rail to a greater height above the tie, and in some cases to hold the ends firmly at the joints. Sometimes, also, the rails are laid upon longitudinal wooden stringers placed upon the ties, or bolted together by iron rods across the track without the use of ties. Fig. 37 shows this sys-

FIG. 37.

tem of construction, without the stringers, the rails being set directly upon the concrete foundation.

Iron ties have been used to a limited extent, and in some cases the rails are set upon chairs resting upon concrete.

The best track from the standpoint of the operation of the railway is probably that formed of ordinary T rails laid directly upon the cross-ties without the use of chairs, in the manner used for steam roads. This form of construction is, however, usually unsuitable for track in streets, as the pavement cannot be laid close against the rail at its upper surface. Where stone or wood blocks are used with T rails it is necessary to cut away the corners of the blocks in order to provide a channel for the wheel-flange. This has a tendency to induce greater wear under heavy traffic. With brick pavements bricks are sometimes moulded of special form, with one corner rounded off, so that they may be set firmly against the rail and still leave room for the wheel-flange. This method has proved fairly satisfactory in some places, but has the disadvantage of leaving a corner of the brick exposed to wear.

In most cases where T rails are employed, the rails are allowed to project above the pavement and form a serious obstruction to the ordinary use of the street. Even where the track is well constructed and

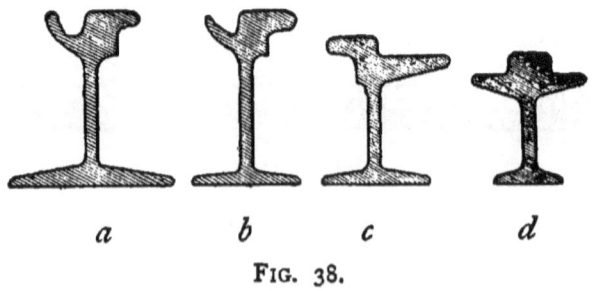

a   *b*   *c*   *d*

FIG. 38.

the pavement originally made even with the top of the rail, under any considerable traffic the wear of the pavement near the rail is usually rapid and the rail soon projects. This is true to a certain extent with any rail, but more especially with the T form.

The form of rail now commonly used in good construction is that known as the girder rail, either the ordinary single web-girder rail as shown in Fig. 38, or the box-girder rail as in Fig. 39.

The advantage these rails possess over the T rail is that the pavement may be laid against the rail, flush with its top surface, the channel for the wheel-flange being provided by the form given to the head of the rail. The box girder is sometimes thought to possess an advantage over the single-web rail from the fact that it affords a vertical surface against which to place the pavement, and an even support to the paving blocks at the bottom as well as at the top, so that there is no tendency for the block to slip under the flange of the

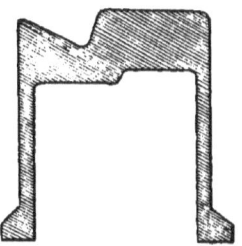

FIG. 39.

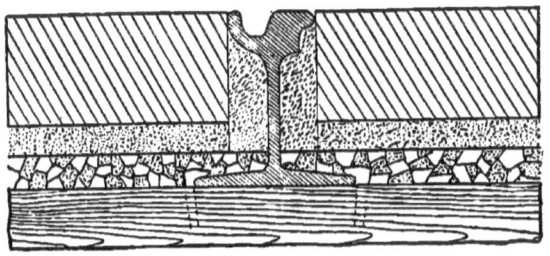

FIG. 40.

rail. In the use of the single-web rails the space under the flanges may with advantage be filled with cement mortar to form a bearing for the paving block as shown in Fig. 40.

Where the paving surface used is not too thick, such as brick or asphalt, the track may usually be constructed by spiking the rails directly to the ties as in Fig. 40. If a thicker surface is to be used, as with a stone-block pavement, the rails must be supported on chairs, unless rails of extra height be used or longitudinal stringers are placed under the rails.

Girder rails, as to the form of head, are divided into centre-bearing, side-bearing, and grooved. Of these the grooved rail of form shown in Fig. 38, a, or Fig. 40, is the most desirable, considered with reference to the ordinary street traffic, and when the pavement is smooth and kept clean is satisfactory in use. It has been extensively used in Washington, D. C. The objection to the use of this form of rail is that the groove is likely to become filled with dirt, and therefore requires constant care to keep clean, especially where the street is not maintained always in good condition. This disadvantage is greater in cold climates where snow and ice are common during winter. It is also necessary with this form of rail that the track be very accurately gauged in width, in order that the flanges may properly fit the grooves; and it is desirable, especially if the rails be supported on chairs, that the rails be tied together by rods as in Fig. 37.

It has been claimed that more power is required to move cars upon rails of this pattern, even under favorable conditions, than is necessary on others. The advantage to the street traffic of using these rails is, however, very considerable. When placed in a smooth pavement which is made flush with the top surface of the rail, the track offers no obstruction to the passing of vehicles over it in any direction, and the inconvenience and difficulty of pulling in and out of the track are avoided.

The grooved rail of form shown in Fig. 38, *b*, is sometimes employed, and obviates to a certain extent the difficulties met in operating track of the form just mentioned, the groove being widened at the top so that the wheel-flange may press the dirt out at the sloping side and also give more room for the flange.

The side-bearing rail as shown in Fig. 38, *c*, is probably more generally used than any other. With this rail the flange extends out on one side to form a channel for the wheel-flange. It is more easily kept clear than the grooved form, but wheels of vehicles readily slip into the channel and leave it with difficulty, although when properly constructed such track offers no resistances to vehicles crossing it. Fig. 41 shows a block pavement with track formed of side-bearing girder rails

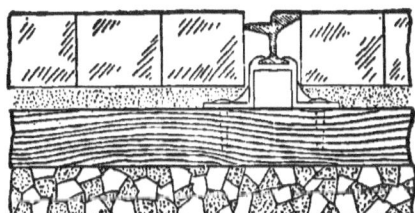

Fig. 41.

supported by chairs which are spiked to the cross-ties.

The centre-bearing rail as shown in Fig. 38, *d*, forms the best track to operate, because it keeps clear of dirt and offers little resistance to the car. It is, however, the most objectionable to the ordinary street traffic, as it is difficult for wheels to cross it; and its use is not commonly permitted on streets of considerable traffic.

Many modifications and combinations of these forms are employed in different localities, and the number of small variations which may be introduced is practically endless. In general, however, nearly all of the rails in common use belong to one of the three classes mentioned.

In addition to the T rails and girder rails various other methods of construction are sometimes employed. The duplex rail is composed of two parts rolled separately and fitting together. The two parts break joints, the object being to eliminate the weakness of the ordinary rail-connection.

Thin strap-rails, or tram-rails as they are commonly called, made to be laid upon longitudinal stringers of wood, are used to some extent, but have in the main been superseded by the girder forms. They consist simply of a plate of iron with a head raised upon it, similar in form to those already mentioned, the plate being laid flat upon the stringer.

The solid construction of track is a matter of importance upon paved streets, because of the difficulty and expense of getting at the track to make repairs, as well as because of the disturbance to traffic when the pavement must be removed for this purpose. The rail-joints and tie-connections are therefore matters requiring particular attention. Where no chairs are used, the use of tie-plates to form a bearing for the rail upon the tie, and to hold it securely in place, is to be recommended, and will greatly aid in forming a rigid track. There are a number of forms in use which give good results. They should be arranged to clamp the rail firmly and present a good bearing upon the tie. When chairs are used, they, like the tie-plates, should clamp the rail firmly and give good bearing surface. They should also be well braced for stiffness against lateral bending.

Joints, in the case of track formed of rails laid directly upon the ties, or upon wooden stringers, are usually made by placing a plate or channel-bar upon each side of the web of the rail ends to be joined and bolting through. The use of slightly curved channel-bars fitting against the flanges of the rail, as shown in Fig. 42, seems to give good results, the spring in the channels serving to prevent the loosening of the bolts.

Where chairs are employed to raise the rails above the ties, joints are frequently most satisfactorily made upon long chairs or bridges reaching across the space between two ties and forming a firm bearing for the ends of the rails.

In order to facilitate keeping the joints tight and enable the bolts at the rail ends to be screwed up without taking up the pavement, joint-boxes are sometimes employed. These consist of openings with removable covers, giving access to the bolts at the ends of the rails.

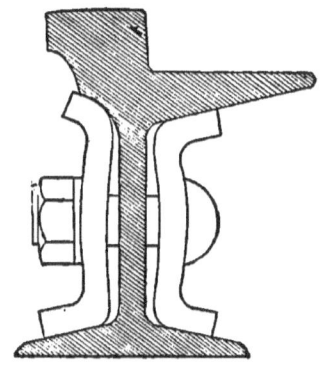

FIG. 42.

It is essential to any good track construction that the track be well ballasted and be brought to an even bearing upon the road-bed; otherwise the track will spring under passing loads and soon become uneven and out of surface with the pavement. Gravel or broken stone is usually preferred for ballast, but where first-class pavements are employed, founded upon a concrete base, the track should also be set in the concrete. This practice has been commonly followed in

Europe with good results. The ballast should be firmly tamped about the ties, which are preferably of hewn timber on account of the greater ease of tamping.

The wear of a pavement is usually considerably increased by railway tracks upon the street. The extent of this wear depends upon the nature of the paving surface as well as upon the construction of the track. It is mainly the difference in resistance to abrasive wear between the rails and the paving surface which causes uneven and more rapid wear of the pavement in vicinity of the track. A broken-stone surface, on account of its rapid wear, is particularly objectionable along a line of track, and is very difficult to keep in proper surface.

Where T-rail construction is used there is a largely increased wear due to the exposed edges of the paving blocks, which wear rapidly on the sides and in the grooves left for the wheel-flanges. (See article by W. L. Dickinson in *Good Roads* for May 1894.) With a smooth pavement and grooved rails the wear is reduced to a minimum where the street is of sufficient width to accommodate the traffic without necessitating the driving of loaded vehicles along the track.

In the case of narrow streets or rough side-pavements the use of the track for hauling heavy loads causes the cutting of the pavement upon the outside of the track, due to the gauge of trucks being greater than that of the track. This is especially the case where, owing to the use of side-bearing or centre-bearing rails, the flange grooves are wide enough to permit the wheels of trucks to enter them.

Where cable roads are used ties are not employed, but the whole structure rests upon the yokes, which pass under the cable conduit and sustain the rails upon their extremities. The conduits are usually built of concrete, which is also used for the base of the pavement, so that the whole structure becomes practically monolithic.

ART. 83. TREES FOR STREETS.

It is always desirable, wherever possible, to have streets, at least those devoted to residential purposes, lined with rows of trees upon each side, both for the purpose of giving shade and to add to the beauty of appearance of the street.

The most satisfactory way of arranging trees is usually to have a tree space between the sidewalk and the curb in which the trees are planted in a straight line along the street. Sometimes in very wide streets a tree space or parking is arranged in the middle of the street, with a driveway on each side. Trees should be spaced in the rows at such distances as will permit each tree when fully grown to spread to its full natural dimensions,—which usually requires, for trees ordinarily employed, from 25 to 40 feet.

The selection of the variety of trees to be used for this purpose must of course depend upon climatic and local conditions. Those which rapidly attain their full size are usually to be preferred. They should have a graceful form and make a good shade, but the foliage should not be too dense. Evergreens are not generally desirable for this purpose. Where there is plenty of room for their development the large-growing varieties with light foliage are handsome and desirable. The size, however, must be suited to the space, and upon narrow streets, or where the trees are to be close to the buildings, they must be of small growth. The ease with which the tree may be grown and its liability to disease or to be affected by the contaminations of a city atmosphere must be considered, as the conditions under which street trees must be grown are not usually favorable to their best development.

It is desirable, especially in cities of considerable size, that the planting and care of trees be under control of the municipal authorities. Trees may then be set with a view to the best general effect upon the street as a whole, the selection and planting of the trees may be properly done, and the trees after planting may be systematically cared for.

Art. 84. Alleys.

The pavements for alleys in cities are constructed in a manner similar to those for streets. Cobblestones, block-stone, brick, and asphalt are commonly employed.

The maintenance of alleys in good condition is a matter of no less importance than the maintenance of streets, although it is more likely to be neglected. It is of special importance that the pavement of an alley be impervious, well drained, and easily cleaned.

The surface drainage of alleys is secured either by forming the section as in a street, with a crown at the middle and gutters and curbs at the sides, or, as is commonly preferable with narrow alleys, by placing the gutter at the middle and sloping the pavement from the sides to the centre. Where the gutter is in the middle it is common to make the bottom of the gutter of a flagstone 15 to 18 inches wide. Fig. 43 shows a centre-drained alley with block-stone pavement upon sand foundation.

Where the pavement is cobble or rough blocks it is

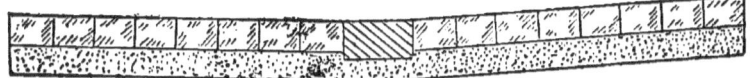

FIG. 43.

desirable also to form side-gutters of flagstones in order to promote ready drainage. Such construction is rep-

FIG. 44.

resented in Fig. 44, which shows a cobble pavement on a gravel base, with curb and narrow sidewalk.

www.ingramcontent.com/pod-product-compliance
Lightning Source LLC
Chambersburg PA
CBHW031811230426
43669CB00009B/1101